Democratizing Data: User-Friendly Visualizations for Particle Physics

Youssva

Democratizing Data: User-Friendly
Visualizations for Particle Physics

Contents

1 Introduction

Particle physics aims to understand the behaviour of matter at the smallest possible scales. Research in this field has historically been split between theorists (who seek mathematical models explaining or predicting observed behaviour) and experimentalists (who seek to produce and detect exotic states of matter that either confirm or confound theory).

The primary tool of the experimentalist is the particle accelerator, which collides beams of massive charged particles together at ever higher energies. These collision events produce showers of new, often exotic, particles. These particles pass through sophisticated detectors, purpose-built electronic devices which register these transits as sequences of electronic signals. The statistical analysis of these signals is the backbone of experimental physics, and a large and powerful software ecosystem is continuously developed and updated to efficiently enable this.

1.1 Problem statement

In the continuing search for new physics, experimentalists have steadily increased both the energy and luminosity[1] of these collisions, creating ever more data. The Large Hadron Collider (LHC) at the European Organisation for Nuclear Research (CERN) is the largest particle accelerator in the world, and produces proton and heavy-ion collisions that are studied by ALICE (A Large Ion Collider Experiment).

Physicists visualise this event data to verify expected behaviour, identify anomalous data, or explain important results. They often use simple visualisations, typically 1 or 2-dimensional histograms as well as line or bar graphs. More complex visualisation is handled by event displays, visual representations of both raw and reconstructed data from a collision.

ALICE is currently being upgraded to significantly increase the quantity of data it records during Run 3 of the LHC. Existing event displays must be modified to support Run 3 data, and none specifically focus on the operation of the Transition Radiation Detector (TRD) within ALICE. These displays often have a steep learning curve, a high barrier to entry, or are tightly bound to a specific environment.

1.2 Aims

The field of visualisation has made great strides in parallel with the data-centric focus of modern innovation. There now exists significant literature presenting approaches, principles and heuristics for building effective visualisations of complex data that clearly communicate important information to the viewer. The design study is one

---

[1] A measure of the particle density of the beam, which is related to the collision rate within the accelerator.

such approach which combines interviews with potential users, and prototype-based discussion and feedback, to design user-centric visualisations for a given domain. We aim to apply this knowledge to design and prototype a simple, portable, user-friendly event display, focused on the TRD, that can support the display of Run 3 data.

The primary audience for this display is experimental physicists with a specific interest in the operation of the TRD and the resultant data. Their familiarity with the physical concepts underlying the operation of the TRD allows for more focused interviews with motivated participants. This audience would also be able to meaningfully critique the physical validity of visual representations, and suggest potential improvements.

The secondary audience is the general public with no knowledge of particle physics, with whom CERN conducts outreach and education activities. Visual appeal, clarity of information and a simple interface are more important than fidelity in such scenarios. This may conflict with the needs of the primary audience, but our expectation is that appropriate design choices can allow the initial single prototype to be easily customised for a different audience. We chose to focus on the primary audience for initial prototypes, with the secondary audience addressed once a robust and functional prototype was complete.

1.3 Approach

We apply a design study methodology to prototype an event display focused on the TRD within ALICE. As pre-requisites we review the fields of particle physics (in order to meaningfully converse with scientists) and visualisation (to identify best-practices to apply). Existing event displays are then critically analysed to identify successful ideas that can be built upon, and shortcomings that can be addressed. We then formulate a methodology based on successful design studies conducted in both physics and related scientific fields.

Our iterative approach involves potential users from a range of backgrounds and expertise. In each iteration, we summarise the themes of in-person interviews to scope and define a series of visual components. We then apply best practices for visualisation from literature to create interactive prototypes that demonstrate their functions. These prototypes are critically evaluated with new and existing users, and their feedback is used to refine the design of the next iteration.

1.4 Contribution

We document here our chosen methodology based around collaborative, user-centric design, as well as some of the use cases for event displays, and the hindrances preventing their adoption and use. We also propose novel data visualisations that provide new ways of understanding raw and reconstructed data from the TRD. These are combined into the primary artefact of this process, a functional and effective event display. A series of case studies then evaluate its use by both scientists and the general public.

A secondary artefact is a generic software tool[2] to prepare both raw and reconstructed data from Run 2 of the LHC for use in the event display. With the resulting documentation and tools, we hope to make the case for a greater use of high quality data visualisation in particle physics, across a broad spectrum of possible users.

1.5 Structure

Chapter 2 is a review of the field of experimental particle physics. We briefly summarise the historical development of the field, before documenting the hardware and software that is central to its modern application. Chapter 3 is a review of the field of visualisation, beginning with a definition of the field and its relevance to experimental particle physics. This is followed by a discussion of guidelines from literature for designing and evaluating visualisations, and a critical evaluation of existing visualisations.

Chapter 4 introduces the design study and reviews successful uses from literature. We then state the expected outcomes of this work, and outline our approach that uses our design study methodology to create a functional, effective, validated event display. The results of the design study are documented in Chapter 5, along with personal reflections on the process. Our conclusions are presented in Chapter 6 along with a discussion of potential directions for future work.

The interview template used in the study is reproduced in Appendix A, and the definitions of physical variables we reference appears in Appendix B. Details of the data formats defined in this work, and the manipulation required to conform to them, appear in Appendix C. Details relating to the implementation of the various prototypes are discussed in Appendix D.

---

[2]Implemented in C++ within the AliRoot framework.

2 Experimental particle physics

2.1 A brief history

The field of physics investigates how our world works. Particle physics focuses on the study of what constitutes matter (the building blocks) and the rules governing its behaviour (the four fundamental forces). The developments discussed below were the result of collaborative work between many scientists, but the figures most publicly associated with each step are named for simplicity.

2.1.1 Building blocks

Over centuries our model of what constitutes matter has been progressively refined by a succession of philosophers and scientists. The original Greek idea of indivisible atoms[1] was ignored for centuries until John Dalton in 1803 proposed that all matter was composed of indivisible atoms, the *solid sphere model*. J.J. Thomson's discovery of electrons (e^-) in 1904 expanded this into a *plum pudding model*, where negatively charged electrons are distributed uniformly through a positively charged sphere. Ernest Rutherford's experiments with alpha particles showed that the positive charge within an atom had to be tightly confined within a central nucleus, leading to the *nuclear model*. Niels Bohr proposed the *planetary model* in 1913, where electrons are confined to specific orbits around the nucleus within quantised energy levels. The work of Erwin Schrödinger and others around 1926 finally led to the modern accepted *quantum model* of the atom, in which the position of electron clouds are described probabilistically as orbitals within an atom.

2.1.2 Forces, fields and mediating particles

Understanding the nature of matter only explains what the world is made of. To understand how it works, we need a model that explains the interactions between matter. Throughout history people have experienced many apparently different types of interactions that were explained in a variety of ways. Our understanding of these interactions developed alongside our deepening understanding of the building blocks of matter. Each proposed explanation sought to generalise disparate behaviours under a single unified theory, a goal which has not yet been achieved.

The current scientific consensus identifies four fundamental forces that are responsible for the state and evolution of our universe over time:

- **Gravity** describes the attractive force between massive objects.

- **Electromagnetism** is a theory that unifies the electric force (which describes the attractive or repulsive force between charged objects) with the magnetic force (which describes the interaction between charged particles in motion).

---

[1] *atom* derives from the Greek *atomos* meaning 'indivisible'.

- The **Weak interaction** describes beta-decay, which allows neutrons to decay into protons (and vice-versa, under certain conditions) and release a beta-particle (electron) in the process.

- The **Strong interaction** explains both, how the component quarks of nucleons in an atom are bound together, as well as how neutrons and protons are bound together in atomic nuclei[2].

The four forces each have an associated field that permeates space and is responsible for the effects associated with that force. Forces and fields are related by mediating particles that transmit a force between objects in a field. The *Standard Model* of physics is a single theory that unifies three of these forces (Electromagnetism, Strong and Weak interactions).

2.1.3 Hierarchy of matter

As discussed previously, our model of the atom has been refined over time into our current picture of a nucleus of protons and neutrons, surrounded by orbital clouds of electrons. This is not the end of the story however. **Quantum mechanics** is the study of matter at the subatomic level, where the Standard Model explains the interactions between *electrons, protons, neutrons*, as well as the atoms formed by these particles.

Quantum electrodynamics introduced the *photon* (γ) to explain the wave-particle duality of light as another fundamental particle, alongside the mediating particles from the fundamental forces. The companion theory of **quantum chromodynamics** has revealed that protons and neutrons are themselves constructed of *quarks*[3] bound together by *gluons*.

This leads us to our current list of elementary particles, from which all other matter is ultimately constructed, as illustrated in Figure 2.1. Combinations of quarks together with binding gluons form hadrons. The most recognisable hadrons are the neutron (one up and two down quarks) and proton (two up and one down quark), which together with the electron (an elementary lepton) form atoms, which we see through photons (an elementary boson). Atoms combine to form molecules of increasing complexity, and both exist in varying states of decreasingly ordered structure (solid, liquid, gas, plasma) to form what we know and experience on earth.

A fifth state of matter existed in the earliest moments of our universe, shortly after the Big Bang. Our best models suggest that this period was dominated by an incredibly hot and dense soup of matter known as the **Quark-Gluon Plasma** (QGP). In this state quarks and gluons are free to move and interact – in contrast to their usual confinement in hadrons – alongside electrons and photons. This deconfinement is a direct consequence of the very high temperature within the QGP ($\sim$ 175 MeV or $\sim 10^{12}$ K [31]), which is more than 100 000 times the internal temperature of the sun. As this fireball cools down, free quarks can bind together into unstable particles that are rarely found in nature. Understanding these particles and their evolution into more stable forms of matter can provide us deeper insights into the early universe.

---

[2] This contrasts with the expectation that the electrostatic forces between a collection of tightly packed, positively charged protons would cause them to violently repel each other.

[3] Which in turn can be differentiated according to: spin; charge; colour; matter vs anti-matter.

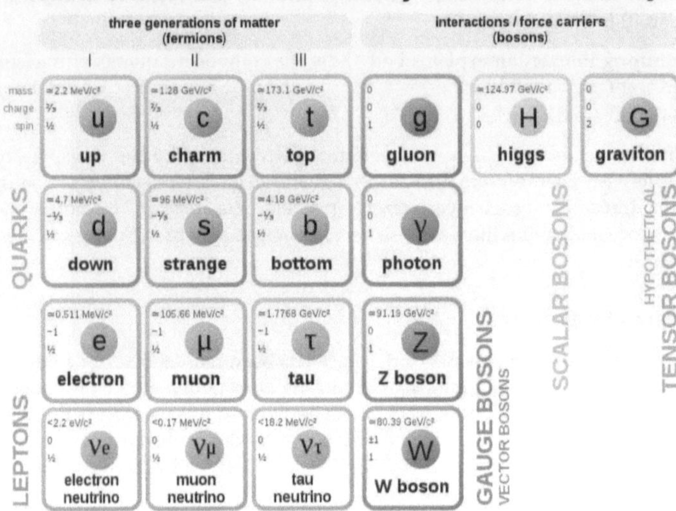

FIGURE 2.1: Categorisation of elementary particles into
Bosons and Fermions [28].

2.1.4 The final frontier

Although the Standard Model has been validated in every experimental test conducted, there remain unanswered questions within the field. The most fundamental of these is the aforementioned search for a unified theory that combines the discrete, probabilistic theory of quantum mechanics at the subatomic scale with the continuous, deterministic theory of special relativity and gravity at the macroscopic scale. The evidence for the existence of dark matter [86] and dark energy [68] raises questions about both their nature and behaviour.

The discovery of the Higgs Boson in 2018 at CERN by the ATLAS and CMS collaborations [1] further validated the predictions of the Standard Model, but also reignited the search for the missing mediators of the gravitational force, the hypothetical *graviton*. More pragmatically, experimental physicists are constantly attempting to increase the precision of our measurements of known fundamental and derived constants. Observing, measuring and understanding particle interactions and the resultant products is the realm of **Particle Physics**.

2.1.5 A need for speed

Particle physics involves observation and measurement at almost unimaginable scales. Atomic nucleus radii are of the order of $\sim 10^{-15}$ m across, while individual quarks are point-like particles with an upper limit on their size of $\sim 10^{-18}$ m. Experimental particle physicists probe the nature of matter in much the same way an overenthusiastic child might seek to understand the contents of a piñata: break it apart with a stick; observe the pieces that remain; and work backwards to deduce the original state.

The piñata in this analogy are the various particles of matter discussed previously, both fundamental (electrons, photons, quarks) and composite (protons, neutrons, atoms). Rather than use a stick, however, they collide bunches of piñatas (particles) together to observe the outcome. This can occur naturally due to the interaction of cosmic rays with the earth's atmosphere, or artificially through the use of **particle accelerators**. The more energetic the collision, the greater the candy bounty – or more prosaically the more numerous and exotic the matter that is ejected from the collision.

Cosmic rays, particles originating from cataclysmic events in deep space, provide a natural source of such collisions. They enter the earth's atmosphere at close to the speed of light, and the subsequent collisions can be many orders of magnitude more energetic ($\sim 1 \times 10^{20}$ eV) than anything we are currently capable of producing in a controlled manner; however these events are very rare and unpredictable.

Particle accelerators are machines constructed to accelerate beams of particles to precisely controlled energies before colliding them in a repeatable manner. The earlier electrostatic accelerators have been superseded by electromagnetic accelerators which are now the mainstay of modern particle physics [75]. Modern accelerators are differentiated primarily by their geometry (linear or circular), the maximum energy they are able to achieve, and their maximum luminosity.

$$L = \frac{1}{\sigma}\frac{dN}{dt} \tag{2.1}$$

The luminosity of a particle accelerator quantifies the rate of events (N) measured in an area over time (t), and is defined by Equation 2.1. σ is the cross-section of the collision, an estimate of the probability of a collision when two beams cross. The integral of luminosity over time (L_{int}) is an important design parameter that all particle accelerators aim to maximise as a way to increase the volume of data generated and subsequently analysed. The LHC reached a maximum luminosity of $2.1 \times 10^{34}\,\mathrm{cm}^{-2}\mathrm{s}^{-1}$ for proton-proton collisions on 5 May 2018 [17].

2.1.6   An invisible assault

Physical principles decree that observation requires interaction, hence it is not possible to directly observe these collisions. Experimental physicists instead rely on **particle detectors** to observe the candy[4] that spills out, and then rewind time to **reconstruct** the sequence of events that occurred in the collision. These observations can then be used to either test an existing theory, or provide evidence for interactions that require the development of new theoretical models.

2.1.6.1   Searching for suspects

Theoretical physicists propose a variety of theories that can describe the evolution of a collision. Experimental physicists then analyse collision data to search for evidence that supports or contradicts theory. The raw collision data from detectors is

---

[4]The particles created in the collision.

reconstructed into a series of tracks, each of which represents a single detected particle. Tracks can then be analysed at varying levels of detail, using a series of filters[5] to focus on aspects of the collision that are deemed important. The most common measurements required for each track are: position; momentum; and identity.

The position of the initial interaction at the start of the collision[6] is important in establishing a reference point relative to which all subsequent interactions occur. Secondary vertices are positions at which produced particles subsequently interact or decay, producing yet more particles. All positions are measured in 3-dimensional coordinates, with an optional fourth coordinate for the time.

The momentum of a particle (p) is a measure of both its inertial (rest) mass (m_0) and its velocity (v), given by:

$$p = \frac{m_0 v}{\sqrt{1 - v^2/c^2}} = \gamma m_0 v$$
$$\gamma = \frac{1}{\sqrt{1 - \beta^2}}$$
$$\beta = \frac{v}{c}$$

(2.2)

where c is the speed of light in a vacuum and γ and β are commonly defined kinematic variables that are useful in relativistic calculations (see Table B.1).

The principle of conservation of momentum guarantees that the sum of momenta for all particles in a given collision (initial and secondary) is zero, in the centre-of-mass reference frame for the collision. Combined with the principle of conservation of energy[7], physicists are then able to determine the invariant mass[8] and velocity of individual particles that are not otherwise identifiable.

The ultimate goal of this model is to identify the particles involved. The electrical charge and invariant mass is generally sufficient to determine precisely what a particle is. This in turn is required to compare and validate the observed evolution of the collision against theoretical expectations.

2.1.6.2 Collision scene investigation

Detectors of different types specialise in measuring different attributes of particles. The choice of detector is motivated by a range of factors, most notably: cost; complexity; and performance. Modern experiments use a variety of detectors, each with their own performance and detection characteristics, to build a more complete picture of the collision and subsequent interactions than any single detector could on its own. Particle detectors are arranged around the point at which the counter-rotating beams intersect, to capture as much information as possible about the interactions that occur, and the particles that are created.

---

[5] Commonly referred to as *cuts* by particle physicists.

[6] Physicists refer to this as the primary interaction or primary vertex.

[7] The initial collision energy is a parameter of the accelerator.

[8] Also known as the *inertial* or *rest* mass.

Many experiments maintain a strong, constant magnetic field in their interior volume parallel to the direction of the particle beams. This field applies a force to charged particles emitted in the collision, perpendicular to the direction of travel. This results in charged particles travelling in a helix, with direction of curvature determined by whether the particle is positively or negatively charged. Neutral particles are unaffected and maintain their straight line path. This effect is an additional source of information for particle identification.

All detectors additionally rely on interactions between particles of interest and the detector itself to conduct measurements [30]. **Gaseous ionisation** detectors rely on the interaction between incident energetic particles and a gas volume, which liberates one or more outer electrons from the gas. **Solid state** silicon detectors exploit the fact that ionising radiation passing through a semiconductor creates electron-hole pairs. In both types, the liberated electrons then drift under the effect of an applied electrostatic field, and are read out as a charge deposition on one end of the detector. These measurements can in turn be used to determine specific observables of interest in the model. The most common observables, and their corresponding measurable quantities, as taken from Sonneveld [80], are:

- **Momentum** (p) can be determined from how much the path of a particle curves under the effect of a magnetic field. This only applies to charged particles, and is quantified as the *bending radius* and *inclination*.

- **Velocity** (v) can be determined by measuring how much time passes between a particle interacting with two different parts of a detector[9] (the time-of-flight). It can also be measured using the emission angle of Cherenkov radiation[10].

- **Charge** (Q) can be determined by the direction in which the trajectory of a particle is deflected. The trajectory of positive and negative charged particles are bent in opposite directions, while neutral particles are unaffected.

- **Lifetime** (τ) applies to the previously mentioned unstable particles, which decay into secondary products after a period of time, which can be measured using the distance travelled before decay.

- **Energy** (E) of a particle can be directly measured using calorimeters, which determine how much energy a particle (neutral or charged) deposits in a medium before coming to rest. It can also be calculated using conservation of momentum and observed momentum values from other particles.

2.1.6.3 Solving a quantum mystery

The signals and measurements recorded by detectors are imperfect clues to the actual sequence of events in a collision, and the signal-to-noise ratio is incredibly low. **Trigger signals** from detectors identify events of interest – the gigabytes of information generated per collision can then be ignored for non-triggered events[11]. **Particle tracking** gleans important information based on the shape and origin of the paths identified particles take through the detector. **Particle identification** then identifies the type and characteristics of the particles that interacted with the detector.

---

[9] You can also use the time between interactions with separate detectors.

[10] Light emitted when a particle moves through a medium at a speed faster than the speed of light in that medium.

[11] The ALICE O^2 strategy is to not use triggers and instead reduce the volume of data recorded per collision.

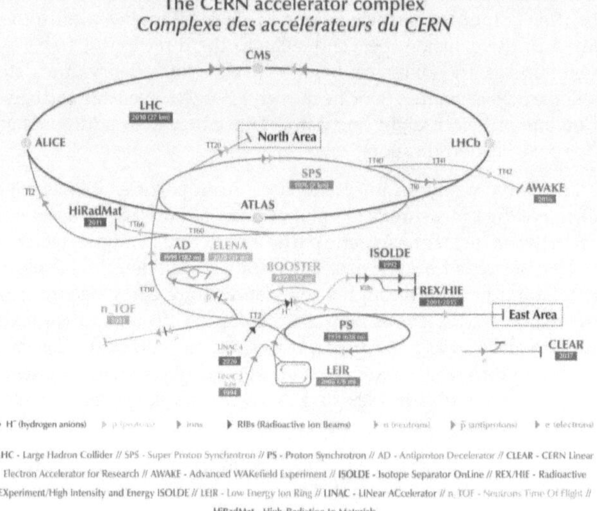

FIGURE 2.2: Schematic view of the Large Hadron Collider at CERN, showing both the main LHC loop and the sequence of smaller accelerators that feed it. The ALICE experiment is visible on the far left.

All this data must then be combined with the expected interactions and predictions of the Standard Model to rewind time and understand where these particles came from, and the sequence of events that led to their production [52]. Only then can experimentalists hope to either find anomalous data (which would indicate the existence of theoretically unexplained physical interactions), or improve the statistical accuracy of accepted values for a variety of universal constants and particle properties.

2.2 Tools of the trade

The frontiers of experimental particle physics can only be explored by reaching ever higher collision energies. This in turn requires financial commitments larger than any single institution or nation can provide. The European Organisation for Nuclear Research (CERN[12]) was formed in 1954 by 12 European states to "assist and encourage the formation of regional research laboratories in order to increase international scientific collaboration..." [43]. In addition to 23 current member states, 8 associated member states and 6 observer states and organisations, CERN also has collaboration agreements with 61 non-member states, a group which South Africa joined in 1992.

Period	Start	End	Status
Run 1	2009	2011	complete
Long Shutdown 1	2012	2014	complete
Run 2	2015	2018	complete
Long Shutdown 2	2019	2020	ongoing
Run 3	2021	2023	proposed

TABLE 2.1: Schedule of operation for the CERN LHC and associated detectors, showing complete, ongoing and proposed phases.

2.2.1 The Large Hadron Collider (LHC)

The LHC is the largest particle accelerator ever built, and is currently the primary focus of experimental physics at CERN. It consists of a 27 km circular, electromagnetic accelerator, fed by a series of smaller accelerators of increasing power, as illustrated in Figure 2.2. The accelerator ring houses two beam pipes in which particles are constrained. Each ferries bunches of particles in opposite directions before colliding at four distinct interaction points on the ring where the beams and bunches cross. Bunches follow 25 ns behind each other[13], resulting in approximately 600 million bunch crossings per second – multiple collisions are possible in each crossing.

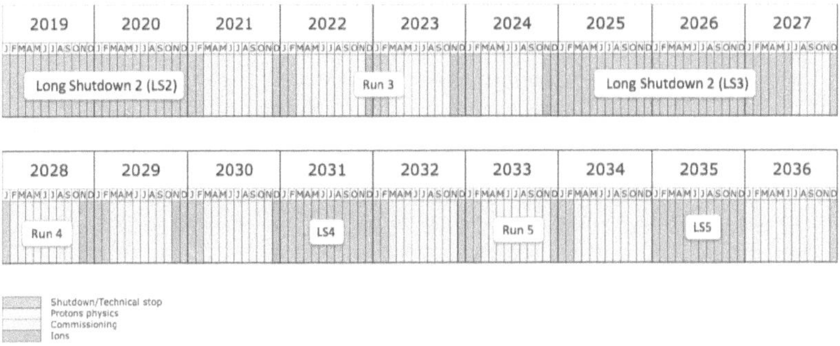

FIGURE 2.3: The long term operational plan for the LHC, showing planned startup (commissioning), proton, heavy-ion, and shutdown operational periods [24].

The LHC is operated continuously for multi-year periods (referred to as Runs), with Long Shutdown (LS) periods between to allow for maintenance, repairs and upgrades, the current schedules for which is shown in Table 2.1 and the proposed future plan in Figure 2.3. During Run 2, the LHC conducted proton-proton collisions for 10 months of the year, with approximately 10^{11} protons in each bunch. For one month proton-lead and lead-lead collisions were conducted where lead nuclei are used instead of protons in one or both rings. These heavy-ion collisions are able to create a quark-gluon plasma [21], the evolution of which produces a variety of particles that can be studied; only a subset of these can be directly detected.

---

[12]The C originally stood for Conseil (council), but this interim body was dissolved when the organisation was established.

[13]The corresponding rate at which bunches cross is 40 MHz.

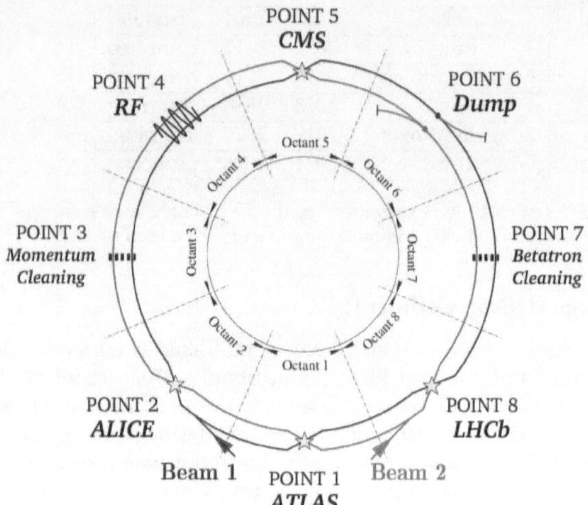

FIGURE 2.4: An illustration of the four interaction points on the LHC ring which house experiments, as well as the counter-rotating beams within the beam-pipe in the ring [44].

The LHC was designed to reach collision energies high enough to explore the quark-gluon plasma. The first collisions were conducted in November 2009 – particles in each of the opposing beams were accelerated to an energy of 3.5 TeV, resulting in a total collision energy of 7 TeV. The most recent collisions were able to reach beam energies of 6.5 TeV, resulting in a total collision energy of 13 TeV. Of the eight interaction points on the LHC ring, four are home to experiments (ALICE, ATLAS, LHCb, CMS), and four are dedicated to beam maintenance and support services[14], as illustrated in Figure 2.4.

2.2.2 A Large Ion Collider Experiment (ALICE)

ALICE (A Large Ion Collider Experiment) is hosted in CERN at point 2 on the LHC ring, and is designed to investigate "the physics of strongly interacting matter and the quark-gluon plasma at extreme values of energy density and temperature in nucleus-nucleus collisions" [9]. ALICE weighs over 10,000 tonnes, is 26 m long, 16 m high, and 16 m wide and sits 56 m below ground close to the village of St Genis-Pouilly in France. The experiment is a collaboration of more than 1000 scientists from over 100 physics institutes in 30 countries [10].

ALICE probes the QGP to compare its observed properties with the expectations of quantum chromodynamics (QCD) which posits the existence of quarks. It also seeks to better understand how quarks are confined in hadrons and to investigate the physics problem of chiral-symmetry restoration. The expansion and cooling of the QGP (freeze-out) leads to the formation of new particles as free quarks bind together, which the detectors in ALICE are designed to observe.

---

[14]High and low-voltage power supply, cooling, and gas supply, among others.

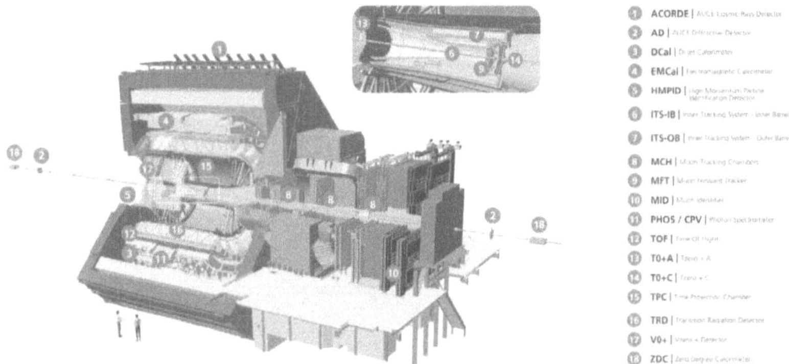

FIGURE 2.5: Schematic view of the ALICE experiment, with annotations for each detector. The TRD is the yellow cylinder (16) around the central blue TPC (15).

Figure 2.5 is a schematic view of the component parts of ALICE, notably: a large, red solenoid magnet surrounding the LHC beam pipe at the interaction point where the beams collide; 18 detectors located within and exterior to this magnet, each optimised for a specific range of particle properties; and support structures, access walkways, and passive absorbers and filters. The solenoid produces a magnetic field parallel to the beam pipe that causes charged particles to follow a curved, helical trajectory as discussed previously.

ALICE uses a right-handed coordinate system as follows: the positive x axis always points towards the centre of the LHC ring; the positive y direction points vertically upward, away from the earth; the positive z direction is therefore tangent to the beam pipe, in an anti-clockwise direction when viewed from above.

ALICE used a trigger-based approach in Run 2, where signals from various detectors are aggregated by the Central Trigger Processor (CTP) to filter and reduce the set of events that were recorded in long-term storage. ALICE is preparing to move to a continuous data acquisition regime in Run 3 [58], where all data is continuously recorded. Together with the planned increase in maximum luminosity the LHC can provide [23] and a change to the beam collision angle in ALICE, this is projected to result in 100 times more data being generated by all the detectors within ALICE. At a readout rate of 50 kHz (which corresponds to data output at 12 GB/s [3]), this necessitates upgrades to both the detectors, physical infrastructure and software during LS2 to handle this increased throughput, alongside the expected maintenance work.

2.2.3 The Transition Radiation Detector (TRD)

The Transition Radiation Detector (TRD) is one of the 18 detectors within ALICE, and is used for event triggering, particle tracking and electron identification. Visualising data related to the TRD is the one of the primary aims of this work.

The TRD is a **multi-wire proportional chamber** (MWPC) with a drift region, that exploits the phenomenon of Transition Radiation to additionally enable discrimination between electrons and other background particles, of which pions are the most common. This discrimination capability is designed to be most effective for fast electrons

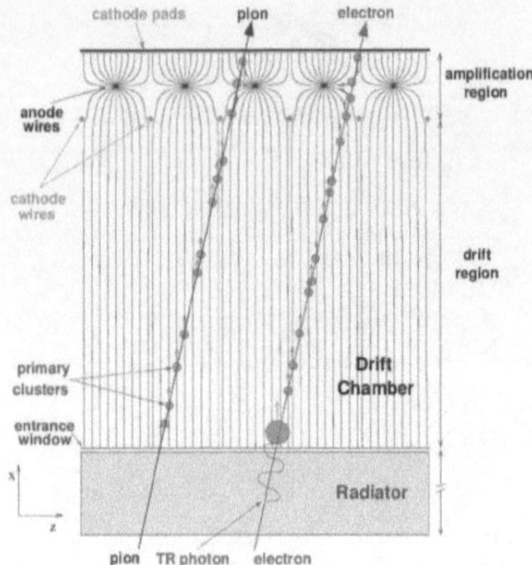

FIGURE 2.6: Cross-section of a single TRD chamber, showing the constituent detector elements and the charge deposition characteristics of pions and electrons [49].

with momenta above 1 GeV, as the primary electron identification capabilities of the TPC[15] detector are less effective in this regime [29]. Electrons in this range are associated with the production of rare, interesting particles like quarkonia and heavy quarks.

The ability to perform identification of charged particles[16] in general, and electrons in particular, is an important contribution to the mission of ALICE to probe the QGP. During freeze-out (cooling) of the QGP, the heavy quarks (bottom and charm) can bind with their corresponding anti-quark to form **quarkonium** (bottomonium and charmonium respectively). The bottom and charm quarks can also bind with any other anti-quark to form **open heavy flavour mesons** (B and D mesons respectively).

The quantity of quarkonium produced is sensitive to the energy density of the QGP, while the quantity of open heavy flavour mesons is sensitive to the initial temperature of the QGP [42]. These products decay to electron/positron pairs with a known probability, and the produced electrons do not interact any further with the QGP. Thus electrons from the primary vertex are an important signal of the QGP, which can be used to study J/ψ suppression and enhanced lepton production in addition to the energy density and temperature of the QGP [87]. The ability of the TRD to separate electrons from the background noise of created particles is therefore very important to the ALICE experiment.

2.2.3.1 Multi-wire proportional chamber

A multi-wire proportional chamber (MWPC) is a type of gaseous ionisation detector that specialises in particle tracking through high-resolution location measurements [26]. The interaction of charged particles (or ionising radiation) with the gas in the detector liberates electrons from the outer-electron-shell of an atom – in the context

---

[15]The Time Projection Chamber detector is the largest detector in ALICE by volume.

[16]Neutral particles neither ionise the gas nor emit transition radiation, and hence are effectively invisible to the TRD.

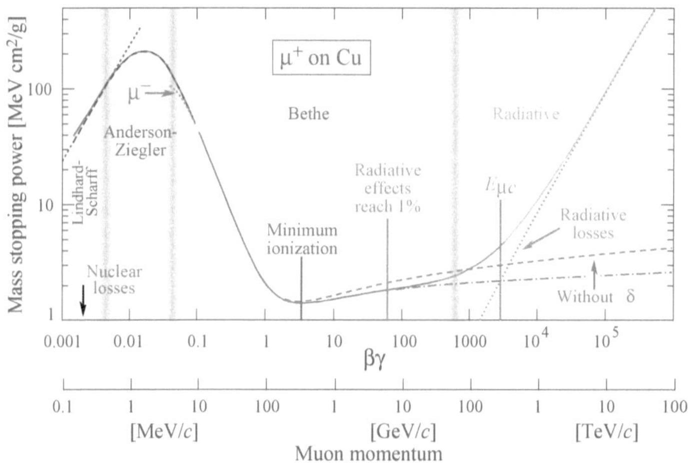

FIGURE 2.7: The Bethe curve showing the specific energy loss for positive muons in copper as a function of momentum ($\beta\gamma = p/Mc$) [83].

of particle physics, this ionising radiation can consist of a wide range of energetic hadrons or leptons.

A MWPC generally consists of an enclosed volume filled with a specifically chosen gas that is both stable and prone to ionisation at relatively low voltages, such as noble gases. Particles created in the collision at the centre of the detector enter through the entrance window and ionise the gas volume before exiting the chamber. This gas volume in the TRD is additionally sandwiched between two electrically charged plates on opposing sides that create a constant electric field perpendicular to the direction of measurement. This creates a drift region adjacent to the MWPC, as illustrated in Figure 2.6.

The passage of particles can ionise the gas, creating an ionisation trail of electrons which then drift under the effect of the electric field towards the cathode and anode wires located on the opposite side. As the electrons approach the anode wires the strong local electric field accelerates them, amplifying the signal by creating a cloud of electrons via an exponential electron cascade. This cloud of electrons is then absorbed by the anode wires, which in turn induces an image charge on the pads in the readout plane that can be measured by the chamber electronics [74]. The MWPC turns these signals into positional tracking of the path of a particle, and the momentum of the particle can then be calculated by determining the curvature of this track.

The energy deposited by a transiting particle in the gas volume through ionisation is governed by the Bethe equation [83]:

$$\left\langle -\frac{dE}{dx} \right\rangle = Kz^2 \frac{Z}{A} \frac{1}{\beta^2} \left[\frac{1}{2} \ln \frac{2m_e c^2 \beta^2 \gamma^2 W_{max}}{I^2} - \beta^2 - \frac{\delta(\beta\gamma)}{2} \right] \qquad (2.3)$$

which relates the rate of energy loss to the atomic number Z and the Lorentz factor

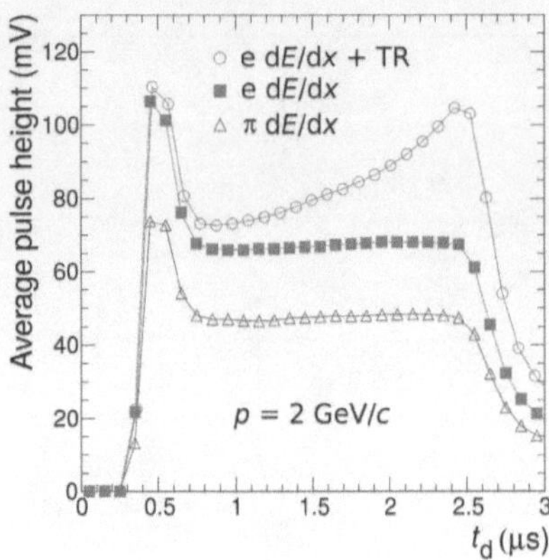

FIGURE 2.8: Plot of the average pulse height (energy deposition) due to pions compared to that due to electrons, both with and without the transition radiation contribution [49].

γ (other variable definitions in Table B.1). This dependence can be used to perform particle identification (PID), but this becomes difficult at high momenta ($>$ 1 GeV) due to the flattening of the curve. An additional PID signal is therefore required.

2.2.3.2 Transition radiation

Transition radiation is the emission of a photon that can occur when an energetic charged particle crosses the boundary between two media with different dielectric constants [5, 37]. The energy deposition due to absorption of this photon in a gas with a high atomic number (Z) is proportional to the Lorentz γ factor, and is therefore large relative to the ionisation energy loss for relativistic particles[17].

This additional deposited energy is overlaid on the background Bethe energy deposition, as shown in Figure 2.8. The particle momentum obtained from the MWPC can then be combined with the γ dependent TR signal to differentiate between electrons and other charged particles at a given momentum, as the light, fast moving electrons deposit more energy than other slow, heavy particles (such as charged pions).

Transition radiation is a low probability event, hence the TRD is designed to increase the probability of occurrence. The radiator is a slab of material, situated before the detector entrance window (Figure 2.6), that provides the medium boundary. In the TRD this material is composed of porous Rohacell and polypropylene fibre mats which increase the number of medium boundary changes, and hence possible photon emissions, to O(100) boundaries in a single radiator. The TRD also layers six chambers into a single module, to further increase the possibility that a transition radiation photon will be both emitted and detected by at least one chamber.

2.2.3.3 Geometry

The TRD uses chambers, illustrated in Figure 2.6, as the lowest level detector element. Each chamber includes multiple rows of 144 cathode pads behind the anode wires to

---

[17]Very fast particles, moving at close to the speed of light.

FIGURE 2.9: The interior of the ALICE detector, opened up for maintenance during Long Shutdown 2. The trapezoidal shapes around the center cavity hold the TRD supermodules, some of which have been removed for upgrading. (S. Perumal, 2019-09-13)

read out the deposited charge (Figure 3.8B), referred to as a **pad row**. Each pad is approximately 7 cm long and 7 mm wide. The 0.5 T magnet that surrounds ALICE causes charged particles produced in collisions to travel in helices (due to the electromagnetic force). The curvature of these helices is most pronounced in the x-y plane, hence the line of readout pads are arranged with the long edge parallel to the beam pipe and the short edge perpendicular to it. This increases the position resolution in the x-y plane.

A **chamber** is a grouping of 12 or 16 pad row chambers[18], five chambers arranged horizontally form a **layer**, and six chambers grouped vertically form a **stack**. The pads in each pad row are tilted by $\pm 2°$ in an alternating pattern relative to the layers above and below (Figure 3.8B); this increases the position resolution in the Z direction. A **supermodule** is constructed of five **stacks** aligned parallel to the beam-pipe (Figure 3.3) and 18 supermodules are arranged radially around the beam-pipe at $20°$ intervals to form the full TRD (Figure 2.9).

2.2.4   Reconstruction

The raw data measurements from the TRD (as with all detectors in ALICE) must be extracted, processed and then analysed, in order to **reconstruct** the sequence of particles created in a given collision. Online processing occurs synchronously while the LHC is operational, and hence is primarily concerned with filtering and compressing the raw data – this is necessary to manage the huge data volumes generated by the TRD and ALICE. Online processing is time-sensitive and therefore seeks to perform only the minimal transformations required to store data for offline processing. Offline processing is an asynchronous operation that is not time-sensitive, and hence occurs at some later point using the output from the online processing.

---

[18]The centre stack has 12 pad rows, all other stacks have 16.

Online processing in the TRD begins with the charge deposition measured on the pads in a pad row, which is translated by an analogue-to-digital converter (ADC) into a numerical count. This is repeated over a number of time-bins (usually 24 or 30), resulting in a 3-dimensional dataset representing the quantity of charge deposited on each pad over time. This can be visualised as a 3-dimensional histogram as in Figure 3.8A.

The quantity of charge deposited, and its shape over time, are used to perform particle identification. This data is also converted via straight-line fit [49] into a **tracklet**. The tracklet represents an approximation to the path of the detected particle through a specific module, with 140 μm resolution in the y-direction and 7cm resolution in the z-direction. This tracklet has a slope correction applied to account for Lorentz drift in the electric field [5]. The intercept and slope of this fitted tracklet, as well as the PID (particle identification value) are output by the front-end electronics (FEE) to the global tracking unit (GTU).

The GTU attempts to combine four or more tracklets into a single particle track, using a straight line fit (Figure 3.1C). This track is saved to file, along with data from other detectors. Offline processing uses these files to combine data from the various detectors in ALICE to produce another output file with information on every reconstructed particle from the original collision. These output files are usually named `AliESDs.root` by convention.

2.2.5 Many hands make light work

Experimental physics at CERN is based on collaboration between large, diverse teams with varying backgrounds. This in turn necessitates tools and infrastructure that scales beyond a single user, machine or computer. The offline and online processing of data from the detectors illustrates this well; both are computationally intensive tasks that require more computing power than a single machine can provide. The volumes of data processed and the nature of the analysis conducted are best suited to server farms running grid computing infrastructure capable of automatically running tasks in parallel.

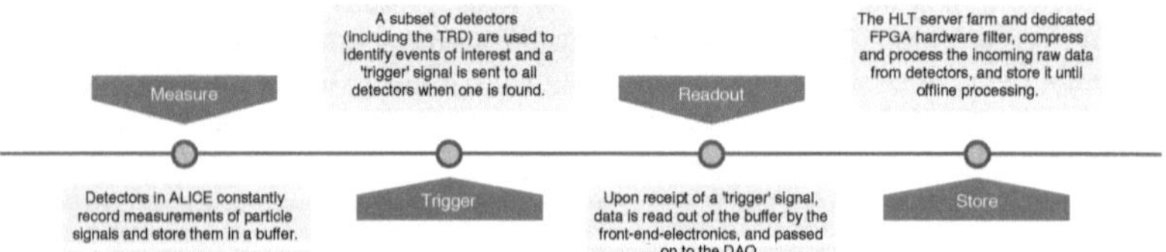

FIGURE 2.10: Visualisation of steps in the online processing of TRD raw data.

Figures 2.10 and 2.11 illustrate a simplified sequence of the steps involved in online and offline processing respectively. The ALICE Data-Acquisition system (DAQ) coordinates and processes the data flow from the detector electronics through to long-term storage. The High-Level Trigger is a server farm that filters, compresses and processes incoming raw data from detectors – it is aided in this task by a collection of customised Field Programmable Gate Array (FPGA) hardware.

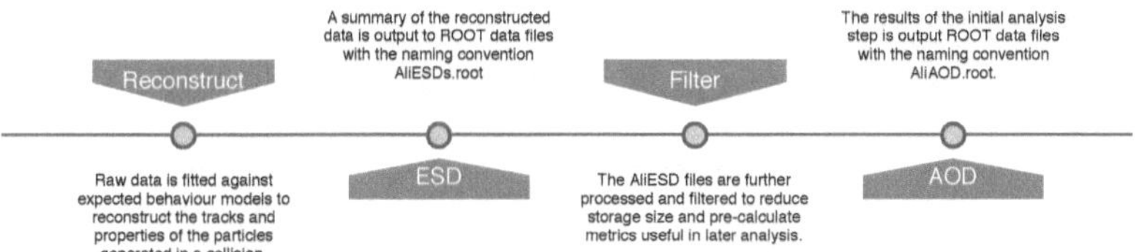

FIGURE 2.11: Visualisation of the steps in the offline reconstruction and analysis of TRD data.

2.3 The digital tool-belt

The primary need within CERN is to perform physics analysis of particle collision data. This can be either real data obtained from the detectors that are part of the LHC, or data from detailed simulations which are used to both verify and predict the behaviour of one or more detectors.

A secondary need is for visualisation tools that allow researchers to derive qualitative intuitions about the data. These are useful in validating the quality of recorded or simulated data, as well as assisting in locating and verifying interesting subsets of data to focus on. The following is a brief review of the analysis and visualisation software that exists and is in use by the ALICE project in CERN.

2.3.1 ROOT

Starting in 1997, CERN developed a tightly integrated software stack in C++ named ROOT [16]. ROOT is the common software environment used for all work within CERN, and supports both static compilation as well as interactive scripting. The current version of ROOT is implemented in C++ 11 and is optimised for parallel access (reading as well as writing) by multiple processes. It is also used by external, non-physics communities for analysis of large datasets [18, 33].

ROOT runs natively on all variants of Unix, as well as MacOS and Windows, and via docker container on other platforms. Initial setup and portability on non-Unix systems is potentially complex, but has been addressed in recent versions.

2.3.2 AliRoot

AliRoot is a customisation of ROOT for the ALICE project within CERN. It is used for event data reconstruction and analysis from Runs 1 and 2 of the ALICE detector [29]. AliRoot benefits from the large library of functionality available within ROOT.

2.3.3 AliEve

AliEve [82] is the existing event display for event data, implemented on top of AliRoot. It is built as an extended C++ class hierarchy that ties together: visualisation data; a GL-renderer; an object editor; and browsable UI elements. The underlying rendering framework does place limitations on the portability of this interface. AliEve relies on the highly-integrated ROOT stack, is natively supported on Linux and MacOS systems and requires Open-GL 3D capable hardware.

AliEve adopts the approach of providing "a versatile selection of algorithms to limit the displayed data to any particular range or region of interest to the user". As such it provides a number of components that can be combined in increasingly complex ways to meet most needs. This toolbox approach does therefore require a degree of technical familiarity on the part of users. Combined with the size and complexity of the rest of the ROOT stack, this can create a large barrier to entry for many classes of users. An example interface under AliEve is illustrated in Figure 3.5.

2.3.4 JSRoot

JSRoot is a full port of the ROOT framework to a web, JavaScript environment [13]. The current version includes sample web based 3-dimensional visualisations of several LHC experiments, together with sample data, as illustrated in Figure 3.7. As a port of ROOT, JSRoot is subject to the same high barrier to entry and complexity of setup for novices found in ROOT.

2.3.5 Run 3 framework and visualisation

As part of the preparation for Run 3 during LS2, the ALICE experiment decided to upgrade their software stack to accommodate the significantly higher processing requirements stemming from the increased luminosity. O^2 is the software framework[19] that replaces AliRoot for Run 3 [19], and is designed for the much higher data throughput rates expected. It is currently being actively developed during LS2.

Part of this development involves a replacement Event Display [55] intended to cover the entire ALICE experiment, whereas this design study is focused on the TRD detector. The outcome of this work can still prove useful in both the initial and ongoing development of the O^2 display, both in terms of the specific requirements around the TRD, as well as more general functional and usability principles that are uncovered.

This O^2 event display is currently under development at the Warsaw University of Technology. The current plan envisions porting the bulk of the existing functionality from AliEve in AliRoot to the new O^2 framework. Progress has been hampered due to several reasons [63]:

- **Incomplete specifications**

  As previously discussed, Run 3 will use a higher overall luminosity, resulting in significantly higher data rates. The format in which this data will be stored, and the internal organisation of the data structures for efficient access are, at present, still in a state of flux. This uncertainty hinders development, and requires cascading code modifications when the data specification changes, due to the tightly coupled nature of framework.

- **Technical debt**

  As a framework that has been built up over decades by a variety of authors, AliRoot and AliEve now represent significant bodies of work. Porting and pruning this codebase is therefore more time intensive than starting from scratch.

---

[19]Pronounced "oh square".

- **Resource availability and desire**

  C++ and ROOT are often not of interest to potential new students. Those who have been approached have expressed an interest in newer, web-based technologies, strengthening the case for a portable, web-based visualisation tool.

- **Implementation of time-frame approach**

  Run 3 data is envisioned to involve continuous data taking. The existing concept of single events, as used in Runs 1 and 2, will no longer exist. It is therefore necessary to develop an interface metaphor to enable users to locate events of interest within time-frame windows. This work is in-development.

- **Availability of data**

  Data formats, and representative data from simulations for Run 3, are difficult to acquire from individual detector teams as much is not yet finalised. This results in delays in prototyping.

A clear opportunity exists to investigate visualisation of Run 2 and Run 3 data using a web based prototype event display that is not based on the existing tech stack. The results of this investigation can inform the ongoing development of official event displays, as well as lay the groundwork for a future move to a more portable tool built for web consumption.

3 Visualising scientific data

This chapter presents an overview of data visualisation using the definitions, approaches and tools discussed in Munzner and Maguire [54]. We first present and discuss the field of visualisation, and the ways in which it can be used to assist scientific enquiry. We then review approaches to designing effective visualisations which adhere to a list of common principles, as well a framework for critically validating and evaluating the resulting designs. We finally present and critique existing visualisations of particle physics, focusing on data and representations that are specific to the ALICE Transition Radiation Detector (TRD), to understand both the problem space and the potential solution space.

3.1 Why visualise at all?

Statistical analysis of data is at the heart of many scientific disciplines. Tabular representations of numerical data can be sufficient to convey well understood information. As data becomes more complex, scientists reach for alternative ways of representing their data. Simple one-dimensional histograms are the most common choice of particle physicists, with higher dimensional histograms and scatter plots also occasionally used[1].

As data volumes increase, computational assistance in generating these representations becomes necessary. Munzner and Maguire [54] describe this as follows: "Computer-based visualization systems provide visual representations of datasets designed to help people carry out tasks more effectively". De Regt [34] similarly states that "visualisation can be an important tool for rendering a theory intelligible", while Sweller [81] discusses how visualisation is imperative to learning and understanding.

Larkin and Simon [51] explore how visualisation can supplement data analysis through diagrammatic representations that expose otherwise implicit information; a particular choice of representation can emphasise features and patterns that are difficult to discern in other representations. Ware [88] defines a *visual query* as the search for patterns in visual displays of data in order to solve a problem.

In general visualisation aims to present quantitative and qualitative existing data in visual forms that facilitate understanding and aid analysis. These forms turn text into graphics that highlight certain aspects of the data, exploiting the powerful parallel processing and pattern matching capabilities of the human visual perception system Ware [89].

---

[1] An unusual combination of all three types is shown in Figure 3.9.

3.2 Applicability to experimental particle physics

Modern experimental particle physics combines theoretical physics, engineering[2], and data analysis. This analysis applies complex domain knowledge to large datasets in massively redundant online database storage systems. The tasks performed can be classified by the clarity of the goal: exploratory analysis involves fuzzily defined tasks looking for patterns; simulation and physical measurements have more clearly defined methods and outputs.

For clear tasks on large datasets, algorithms can automate the required task without visualisation, whereas tasks with small or poorly defined datasets are unlikely to benefit from the effort required to visualise them. There exists a large middle ground between these two extremes, however, where computer-based visualisation can assist scientists in their research. Visualisations can be borrowed from work in seemingly unrelated fields, or developed *sui generis* for specific needs. In either case, computer-based visualisation has the advantage that it can be easily updated or applied to different, often unrelated, datasets with minimum effort.

The human visual system excels at processing multiple visual information streams in parallel [89], and finding patterns in what is seen. Physicists can exploit this to search rapidly for patterns or anomalous data. This is particularly useful in validating the operation of the detector, finding errors in the data, or debugging processing logic alongside routine code analysis.

A well designed visualisation can assist users and researchers who are new to the field by providing much needed context for the available raw data. It can also highlight important aspects of the data while hiding complexity that is not initially relevant, thereby bridging the gap between data and understanding. The most common example of this is the display of reconstructed particle tracks, which hides the complexity of reconstructing these tracks from the raw data measured in a detector.

Visualisations can group together otherwise disparate data for localised display and interpretation, allowing users to focus on the task at hand rather than complexities of data format or storage, as demonstrated by the event displays discussed later in this chapter. These displays use interactive synchronised views of multiple facets of the same dataset, which supports answering multiple visual queries in parallel. Connections and relationships between the separate data views can be highlighted, reducing the mental load imposed on the user as less data needs to be held in memory in order to make sense of the whole.

Interactivity can also manage levels of detail within data. An initial overview of the data can be supplemented by the ability to drill-down on specific areas, revealing detail as required rather than overwhelming the user with the full complexity upfront.

3.3 Designing visualisations

The potentially complex process of design is simplified by incorporating successful approaches from literature which formalise the trade-offs involved, guide the choice of solution, and provide specific rules to apply.

---

[2]The design, development and operation of accelerators, detectors, and related infrastructure.

3.3.1   Axes of design

There are three independent trade-offs that must be managed when designing visualisations. The first is between a specific design that supports a limited set of data and tasks, and a more general design that supports many different goals and datasets. Two additional trade-offs are formulated by Agrawala, Li, and Berthouzoz [4]: detailed versus simplified; and realistic versus abstract presentation.

Generality

A specific design avoids overloading the user with too many initial choices and options. This promotes manageability of both the design and implementation processes at the cost of limited usability and lifespan. This is true of many illustrations in academic papers, as shown in Figure 3.1.

A general design that is flexible and choice-driven can support multiple tasks and user roles, but often necessitates a high initial learning curve which can result in users making ineffective choices without sufficient experience. The AliEve event display in Figure 3.5 is one such example.

Realism

Realistic visualisations closely resemble the actual state or complexity of a system and its data. This has the advantage that individual elements are easy to recognise and contextualise, but potentially hard to reason about as the realistic appearance creates visual clutter that can hide important details. Figure 3.1A is an accurate representation of a detector, but any data overlay would be difficult to discern against the background imagery.

An abstract visualisation aims to capture the essence of a domain by omitting unnecessary detail in favour of an abstract model of the domain. This can prove problematic if the abstraction omits important information necessary to answer a specific visual query. This is evident in Figure 3.6 where different types of recorded and reconstructed data are displayed, but it is very difficult to relate the data to specific parts of the detector.

Simplicity

Detailed visualisations aim to include as many channels of information as possible, enabling a wide range of visual queries to be answered. Such a design can quickly become cluttered and unwieldy, however, especially if sufficient care is not taken to keep the channels visually distinct and separable. The control room interface in Figure 3.4 has very high information density, but requires training and experience to utilise effectively.

Simplified visualisations focus on a subset of information channels to emphasise, and are therefore simpler and quicker to visually parse, but are by design limited in the queries that can be answered. Figure 3.8A focuses on the relationship between raw and reconstructed tracklet data, and excludes other unrelated information.

A combination of several faceted simplified visualisations, bound together via interactive zoom and filter, is one option that combines the best of both approaches, as utilised in the AliEve display in Figure 3.5. The value of this must be balanced against the complexity introduced by the required interaction.

3.3.2 Navigating the solution space

Deciding on an appropriate and effective visualisation solution is a difficult problem, dependent in large part on the familiarity of the designer with a wide range of possible solutions. Munzner and Maguire [54] advocate the approach discussed below where designers: familiarise themselves with as many existing solutions as possible; iteratively winnow down the options under consideration; propose a subset of those for user evaluation; and select a final solution.

A large known search space is achieved by critically analysing existing solutions, as we do in Section 3.5, as well as applying the what-why-how trio technique developed by Munzner and Maguire [54]. This technique breaks visualisation choice down into three questions: **What** is the data being examined? **Why** is the data important? **How** should the data be represented? This trio of questions can be chained together to analyse more complex problems by starting at an abstract, high level, and progressively drilling down into more detailed tasks.

3.3.3 Visualisation principles

It is imperative to emphasise that visualisation as a scientific discipline aims to convey facts and information that support scientific enquiries. There are a multitude of ways to encode information in visual information channels, and designers must identify and exclude ineffective design choices early on. To that end, the following is a list of visualisation principles from Munzner and Maguire [54] that can assist in this selection process, which they motivate with both experimental quantitative and qualitative evidence.

- **No unjustified 3D**: 3-dimensional visualisations are visually very appealing and are very effective for shape and orientation perception, but there are significant drawbacks to their use. The presentation and perception of depth and perspective can distort meaning and lead to incorrect interpretation of data; the human visual system perceives information in the horizontal and vertical planes with much higher density and fidelity than is the case with depth. Objects in the same visual plane can occlude each other, hiding information, and care must be taken to ensure text is not tilted relative to the viewer lest it become illegible.

- **No unjustified 2D**: The use of 2-dimensional visualisations should be explicitly justified over a simple list or table. These 1-dimensional alternatives are optimal for displaying multiple attributes in a given space, and perform very well for lookup tasks with appropriate sorting.

- **Eyes beat memory**: simultaneously visible views impose lower cognitive load than comparing against a view held in memory, due to the limited capacity of our working memory. Animated transitions between different views of the same dataset can help users maintain context and aid understanding. Animating changes in the content of a view can conversely make keeping track of changes difficult, due to the need to keep multiple frames of animation in working memory at the same time. The human visual system is also prone to change blindness, where a focus on changes in one area blinds one to changes in others [78].

- **Resolution over immersion**: Immersion refers to a focus on realism and fidelity. Highly immersive visualisations sacrifice resolution and data display density. Immersion contradicts the principle of *"overview first, zoom and filter, details on demand"* proposed by Shneiderman [77]. This advocates providing additional detail only when requested, after the user has located an area of interest from a high level overview. This ensures related information is displayed close together, maximising contextual awareness on the part of the user.

- **Responsiveness is required**: Interaction allows exploring a larger space than would be possible with a static image, though this comes at the cost of increased cognitive load on the user. Latency can increase this load further by increasing the time between a user initiating an interaction and the result being available and visible. It is therefore important that such operations be as performant and low-latency as possible.

- **Function first, form next**: Goguen and Harrell [41] argue it is more important to preserve structure than content, and hence visualisations should aspire first to be effective and functional in enabling users' visual tasks and queries. Aesthetics are important, particularly given human preference for beauty and visual cohesion, but should be a subsidiary concern, except where they advance perception and effectiveness of communication.

3.4 Validation and evaluation

Evaluating a visualisation, and validating that it addresses user requirements, can be difficult due to the need to balance multiple conflicting concerns. There are many possible metrics that might be relevant, including but not limited to: effectiveness of the visualisation as a whole; the appropriateness of selected visual idioms; quantitative measures of increases in efficiency when performing selected tasks; and qualitative measures of engagement and understanding facilitated by a visual representation. The metrics themselves are also often imprecisely defined and domain specific, and different evaluation strategies are necessary for different audiences.

Our chosen approach uses the framework presented in Munzner and Maguire [54], which begins by defining complementary design and evaluation hierarchies. The framework posits that there are four nested levels of design and validation in any visualisation, and higher levels should be addressed before proceeding to lower levels. In this work we have performed design and validation at all four levels, the details of which are covered in subsequent chapters.

The **domain situation** level is concerned with understanding the skills, experience and needs of potential users within the domain. The primary threat to validity at this level is that the wrong problem is solved during the design phase, resulting in a visualisation that does not address the users' needs. This threat can be mitigated by observing and interviewing potential users at the start of the process, before solution design begins. Our specific strategy for tackling this level is documented in Section 5.1.

The **data and task abstraction** level aims to ensure that the appropriate data is presented in order to facilitate the set of identified tasks. The potential threat during the design phase is that inappropriate data is used, or that required tasks are incorrectly identified or abstracted, both of which can be mitigated in several ways. The *what/why* questions are directly applicable to both design and validation at this level, and applying them correctly can pre-empt many potential issues. Visualisation researchers can additionally test proposed solutions on potential users to collect "anecdotal evidence of utility", or conduct a field study to document how existing systems are utilised. We address this by focusing on the needs of expert users, who are able to accurately identify both the data and tasks required.

The **encoding and interaction** level corresponds to the *how* question. At this level the threat to validity is that an inappropriate or ineffective encoding or idiom is selected to represent the data. This threat can be mitigated by considering all potential channels during design, and then explicitly justifying a selected subset. Visualisation researchers can collect quantitative or qualitative measures of usability and effectiveness through informal interactive studies with users, the approach we have opted for. Alternatively they can conduct more rigorous lab studies to measure elapsed time and average errors per task, both with and without the visualisation solution.

The **algorithmic layer** concerns the implementation of a visual solution based on answers to the *what-why-how* trio of design questions. The primary threat at this layer is a slow or unresponsive interface that negatively impacts a user's experience with an otherwise well designed visualisation. This threat can be addressed by estimating the algorithmic complexity of potential approaches before settling on an implementation. One can also measure and iteratively optimise the consumption of available computing resources (time, memory, cpu utilisation) within an implemented prototype. This level is often omitted in purely design focused studies. In this work we have implemented a full, functioning prototype optimised for responsiveness, discussed in greater detail in Chapter 5.

3.5 Review of existing particle data visualisations

Static and interactive visualisations are used across CERN for a variety of purposes. Some are custom generated for outreach or marketing purposes to the general public. Others are used by experts within the field to communicate important results, interpret recorded data from detectors, or troubleshoot erroneous measurements. Whether designed primarily for visual appeal or scientific enquiry, there are elements within each that could inform the design of an interactive event display for TRD data. This Section summarises the main results of our extensive review of existing visualisations.

3.5.1 Particle trajectories

Our review of existing visualisations and literature reveal that physicists are primarily interested in understanding the trajectory[3] of particles produced in a collision. This is best achieved with a cross-sectional, orthographic view with the viewing plane perpendicular to the beam direction – Figure 3.1 represents four different approaches to rendering this view.

Figure 3.1A is a diagrammatic cross-section of the ALICE experiment and detectors, taken perpendicular to the beam pipe. It is a standard image, drawn from the technical specification of the TRD, used to illustrate the construction and components of ALICE in general and the TRD in particular. This visualisation is highly realistic in its representation of the structural detail and layout of the experiment, but the visual clutter that ensues makes it very difficult to overlay experimental data in a usable way.

Figure 3.1B by contrast is more data focused, and includes a length scale to better contextualise the magnitude of features. Detectors have been reduced to simple colour coded shapes, with particle track data overlaid. Strong use of colour makes discernment of individual detectors easy, but removes colour as an information channel for event data. The component parts of each detector are also abstracted away, making it

---

[3]Helical for charged particles, straight lines for neutral particles.

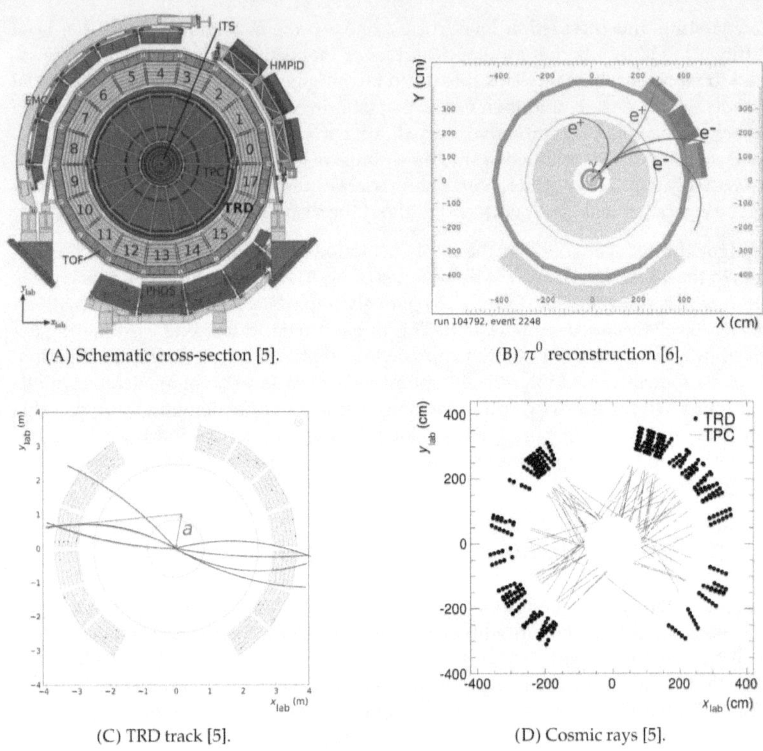

(A) Schematic cross-section [5].

(B) π^0 reconstruction [6].

(C) TRD track [5].

(D) Cosmic rays [5].

FIGURE 3.1:
Comparison of cross-sectional views of the ALICE experiment and TRD detector: (A) a schematic cross-section of the ALICE detector; (B) a reconstruction of a π^0 candidate event; (C) a reconstruction of TRD track (green) using straight line fit, compared with actual helical tracks (blue); (D) a display of reconstructed tracks and tracklets created by cosmic rays [6, 5].

difficult to examine the detailed interaction between a single particle and a particular detector.

Figure 3.1C is a further simplified representation where all detectors have been omitted except the TRD. The TRD modules are illustrated with simple outlines, allowing the raw and reconstructed data to be clearly discerned. This is more effective for the reconstructed tracks, as the tiny tracklet dots are difficult to identify. In contrast to the previous two examples, colour has been reserved as an information channel for event data.

Figure 3.1D takes simplification to the extreme by omitting detectors entirely. This view clearly illustrates the raw data measured by the TRD and TPC detectors, with visually distinct marks and intensities used for each. Despite there being no explicit rendering of the detectors, the density of data makes the location and separation of the detectors easy to infer. This view is effective despite the lack of colour, and provides a

FIGURE 3.2: Display of a side-on view of Pb-Pb collision event [8].

good high level view of the data.

The most common companion view to Figure 3.1 is also an orthographic projection, but of the plane parallel to the beam-line. This view is generally considered less useful as there is limited variation in track characteristics, as evidenced in Figure 3.2. One issue with this view is that the cylindrical nature of the detector allows any given track to cross multiple planes, effectively forming a solid angle that must be projected into 2D. This is generally accounted for by projecting every track onto a single ideal plane, and then rotating that plane to be perpendicular to the viewing plane. Figure 3.2 illustrates a version of this where tracks are projected above and below the beam line depending on their location, allowing them to be compared against each other.

Figure 3.3 shows a side-on view of a single TRD supermodule; 18 of these supermodules are arranged in a cylinder around the beam pipe as shown in Figure 3.4. This view makes clear the differences in dimensions and positions of the modules that form a supermodule. It also contextualises the standard terms Stack and Layer, as well as A-side and C-side which are used to indicate orientation. When contrasted with the similar views in Figures 3.2 and 3.5 it is clear that the discrepancy in size between the TRD and the overall ALICE experiment means that much of the TRD-specific detail is lost if only the overview is considered.

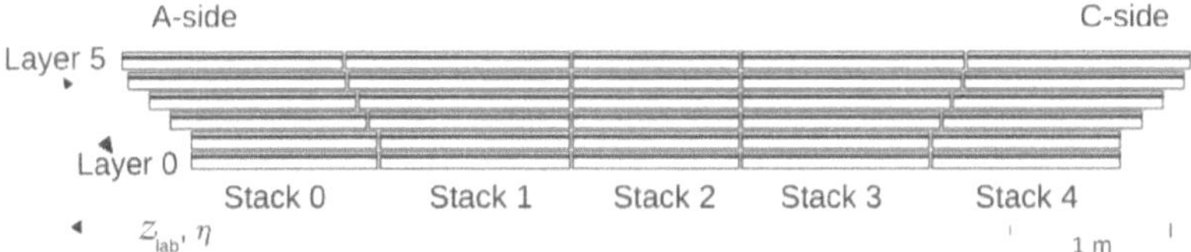

FIGURE 3.3: Side-on view of layer and stack arrangement within a
single supermodule. [5]

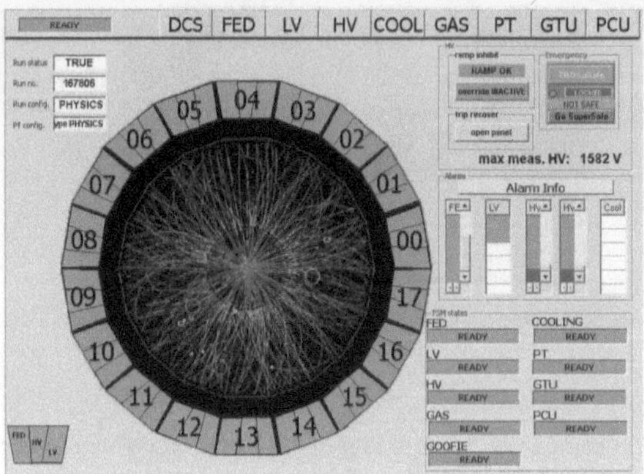

FIGURE 3.4: Graphical user interface used in the ALICE control room
to view and control the state of the TRD detector [5].

3.5.2 Multiple views

The complexity of data produced by ALICE often necessitates faceting of views to contextualise and manage this information. Figure 3.4 is the detector control system (DCS) GUI[4] used in the ALICE control room to monitor all aspects of the TRD [20]. Alongside the now familiar sector views are a large number of visual status indicators and numerical readouts, both using colour to indicate state.

The text abbreviations along the top edge of the display are tabs that when clicked present similar views of different aspects of the detector. The primary points of interest here are a very high density of information and a busy, interactive interface that, with training, is very informative to expert users. There is a steep initial learning curve, however, and the implementation in WinCC[5] reduces the portability of this display to other uses.

Figure 3.5 is a fairly typical setup of the AliEve software. As discussed previously, this is the standard event display in AliRoot (see Section 2.3.2) that is highly configurable, but with a steep initial learning curve and limited portability. Of note here is that there are multiple synchronised, faceted views with a tree-based navigation control on the left. The two previously discussed stack and sector views occupy the right of the display, with an interactive 3D render of the data taking up the main centre area.

Colour is once again used to encode detector identity, with lines and dots used as marks to indicate reconstructed and raw data respectively. As configured in the image, the display is focused on a high level overview of a single event. Tabs are again used to provide multiple views of the data, which conflicts with the "eyes beat memory" principle. Filtering of visible data is possible through the panels on the left of the display.

---

[4]Graphical User Interface.

[5]A custom framework for developing monitoring and control interfaces, an example of which is discussed by Normanyo, Husinu, and Agyare [56].

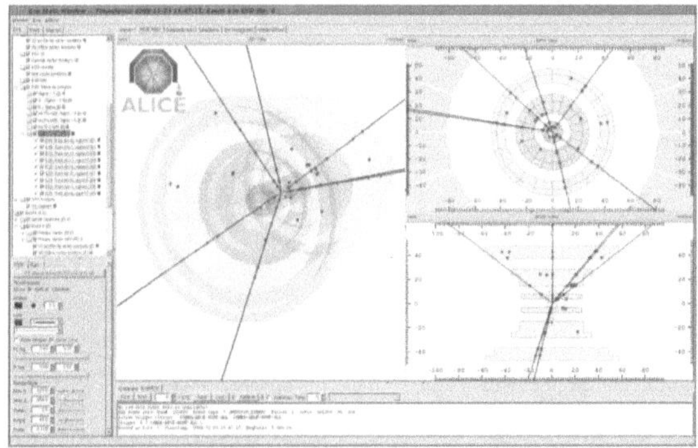

FIGURE 3.5: Multi-panel display of an event in AliEve [2].

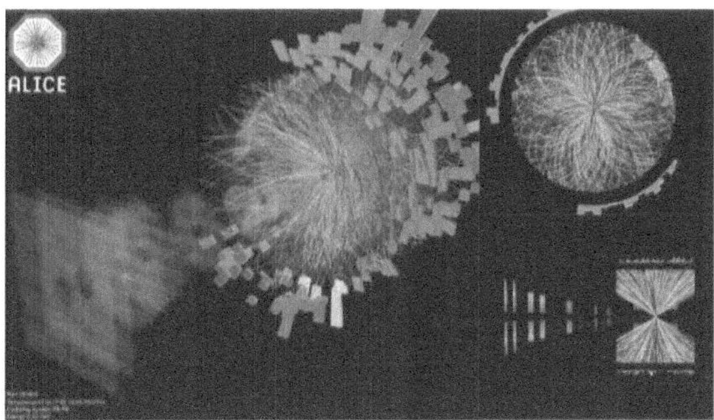

FIGURE 3.6: Pre-rendered illustration of Pb-ion collision, showing three different perspectives on the event [7].

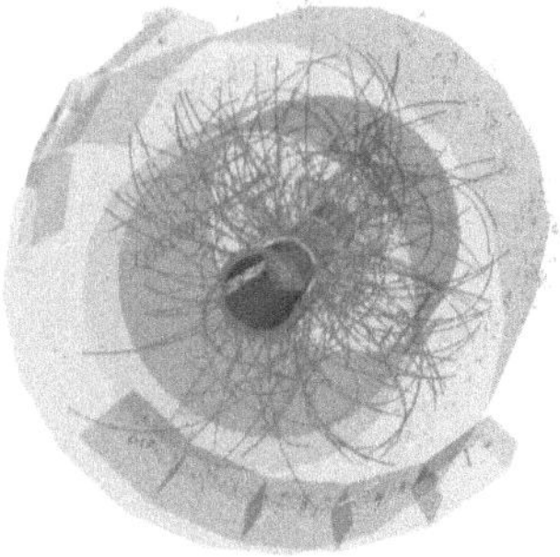

FIGURE 3.7: Display of ALICE in JSRoot, including detectors, hits, and tracks [13].

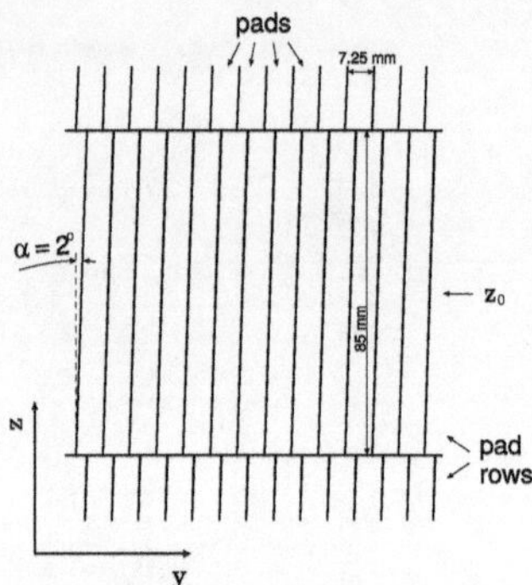

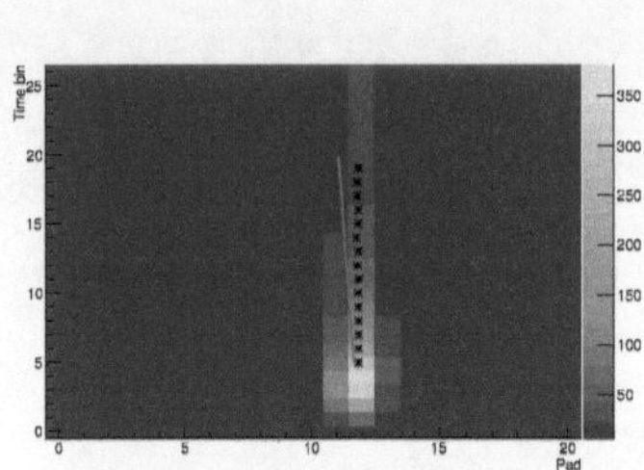

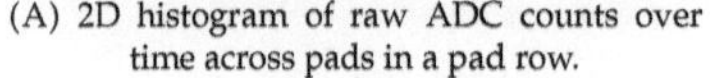

(A) 2D histogram of raw ADC counts over time across pads in a pad row.

(B) Illustration of tilt of pad rows in order to increase the effective z-resolution of the data.

FIGURE 3.8: Two approaches to visualising pads and pad ADC data [5].

3.5.3 The third dimension

3D visualisations have become more widely used as technology has progressed, processing power has increased and tools for implementation have matured. Figure 3.6 is one such visualisation created explicitly for outreach and marketing purposes by CERN. The "no unjustified 3D" principle is addressed in this case by the need to focus on aesthetics and context first, as the intended audience is not expected to have any familiarity with the specifics of the experiment. The overloading of colour for multiple independent information channels (particle energy, detector identity), lack of a key explaining the encodings used, and overwhelming density of information make it clear that this a visualisation meant to be enjoyed, rather than as a tool to investigate data.

Figures 3.5 and 3.7 are similar examples taken from ROOT and JsRoot respectively for use by expert users. These are much more restrained iterations on the same theme, but still overload the colour channel unnecessarily. A common issue here is that the effect of perspective transformations, and the use of opacity to make detector elements partly transparent, combine to obscure salient features of the data.

In use it also becomes apparent that multiple manipulations are required to find an appropriate view of a given track and track selection is difficult due to multiple obscuring elements. Detailed interrogation of the interaction between a particle track and a sub-detector requires constant switching between zoom levels and rotations, again violating the "eyes beat memory" principle. When contrasted with the behaviour and use of the stack and sector 2D projections discussed earlier, it becomes clear that these uses of 3D are not well suited to scientific interrogation.

3.5.4 Raw data

The visualisations discussed thus far have all been at a high level, showing post-reconstruction data for the ALICE experiment as a whole. At an individual detector level, the raw data and experimentalist concerns can look quite different. The two visualisations in Figure 3.8 are specific to the TRD, and illustrate this point well.

Figure 3.8A shows a typical data visualisation of the raw ADC counts for a subset of pads in a single pad-row over a series of time-bins[6]. This histogram[7] illustrates the evolution of the measured signal response due to the transit of a particle through the chamber, and is particularly useful to experimental physicists in monitoring and debugging the performance of the TRD.

Figure 3.8B is also a visualisation of pads in a pad-row[8], but here is a schematic illustration of the physical layout of pads within a pad-row. In addition to showing the actual physical dimensions of a pad, this also reflects the slight $2°$ tilt of the pads to increase the position resolution in the y-direction. This subtle effect is important when working closely with TRD data, but is effectively hidden from upstream consumers by the data pre-processing that occurs during data acquisition by the TRD, and is discussed further in Appendix D.

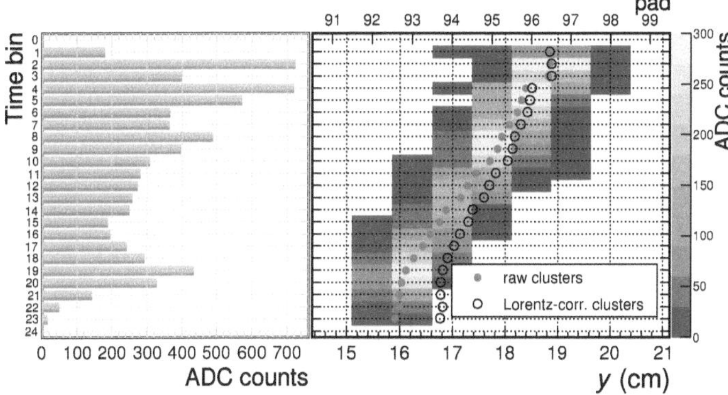

FIGURE 3.9: Dual view of raw TRD data per time-bin over pad rows, with aggregation per time-bin over pads [5].

Figure 3.9 is a novel static visualisation that combines the 2D histogram from 3.8A with a cumulative histogram per time-bin over a subset of pads within a pad-row. A series of points indicating the corrected and raw clusters used to fit a tracklet to this data are overlaid on the histogram.

This combined view allows an experimentalist to understand the evolution of the entire signal deposited by a transiting particle over time. They are also able to validate and troubleshoot the tracklet fitting operation, one of the most important functions of the TRD. The side-by-side view emphasises the "eyes beat memory" principle, and while the colour channel is used for two different information channels, this use is supplemented by the choice of non-conflicting marks. The use of a colour gradient to encode height allows for a 2D rather than 3D representation, abiding by the "no unjustified 3D principle".

---

[6]Section 2.2.4

[7]Displayed in 2D with a colour scale for the third, quantitative dimension.

[8]The use of pads and their layout is discussed in Section 2.2.3.

3.5.5 Summary

Overall there are several common aspects across the reviewed visualisations that could be pulled together into a single cohesive display, with a similar approach to AliEve but requiring less upfront configuration. There are many opportunities to apply the principles of visualisation to improve and expand upon the static visuals. Across the visualisations, colour is used as the primary information channel, while other marks and channels are not widely utilised; exploiting the full range of available idioms would improve information presentation. It is also clear that the existing user-base is comfortable using a variety of visual idioms to navigate the data, suggesting that one or more of the static visuals could be made more dynamic and interactive, so as to convey more information.

4 Methodology: a plan of attack

In this chapter we introduce the Design Study as a framework for designing interactive visualisation solutions. We then discuss when they are appropriate to use, and present approaches and tools that can be used as part of the process. The experiences from existing examples of successful design studies are summarised, and used to define our chosen methodology. We then discuss the anticipated outcomes and artefacts of the process, including the use of case studies and presentation of the prototype at CERN Open Days. Finally we describe our iterative design approach, and detail the steps involved.

4.1 A review of design theory

The effectiveness of interactive scientific visualisations, discussed in Chapter 3, is very dependent on context, intended use and the correct application of visual design principles. A formal process that takes these considerations into account can result in an interactive visualisation that appropriately applies design principles to achieve a contextually useful result. Design studies are a popular approach to achieving this by conducting domain specific visualisation research and specification, as outlined in the work of Sedlmair, Meyer, and Munzner [76].

A design study is defined by Lam, Tory, and Munzner [50] as the process used to develop a tool that addresses a real-world problem using real data, and involves at least one target user in design or evaluation of the tool in a non-trivial way. They present a framework including nine analysis goals to help guide and structure a design study.

4.1.1 Design studies for scientists

Quantitative scientific research is often characterised by domain experts with a deep understanding of the problem at hand, applying domain-specific knowledge and tools to analyse large, potentially complex datasets. Their contextual familiarity with both the data and applied analytical methods allows them to directly extract and interpret meaningful results directly from the available data. While the value of visualisation is widely acknowledged, scientists are focused on solving their specific domain problems. The design and utilisation of visualisations therefore tends to be ad-hoc in nature, for a specific purpose, and customised to the needs of a single scientist or team.

Scientists appreciate the value of tools that provide a re-usable visualisation solution, but they often do not have time or resources to focus exclusively on tool development (see Section 5.1.1). Outsourcing the development of such tools to computer science or software development experts can result in the best software outcome, as they can apply industry best practices and techniques. Effective communication is required to bridge the large gap that can exist between scientists and developers, in terms of both domain knowledge as well as requirements for functionality and interactivity.

A design study brings in an objective outsider [76] to elicit and translate the requirements and expectations of domain experts into a design specification that is clear and explicit enough to enable direct implementation by dedicated software developers. The study should aim to document the process of design, as well as the specific experiences and lessons distilled from working with a range of interested potential users. The results of this cross-domain collaboration can assist future work in developing visualisations in a variety of fields.

4.1.2 The designers toolbox

The field of design is broad and there are many complementary methodologies that can be applied to a given design problem. The work of [76] reviews a number of design studies that have been conducted across a variety of problem domains, and serves as the primary guiding source for what follows. In this work, they summarise the following generic steps common to all successful design studies, which attempt to address the four threats to validity identified in [53]:

1. Analysing the problem to ensure that the correct problem is being addressed and all parties share a common understanding of this problem.

2. Identifying and abstracting the tasks associated with this problem, to ensure that proposed visualisations present the correct data in a manner that supports the identified tasks.

3. Designing and implementing an effective visualisation solution that applies the principles of visualisation (discussed in Section 3.3.3).

4. Evaluating the proposed solution with real users to validate the design and identify areas of future improvement.

5. Documenting the design process, findings and proposed solution.

Below we present a number of specific techniques that were considered for use in this design study, and discuss their contextual strengths and weaknesses.

4.1.2.1 Interviews

At the outset of the design study, it is necessary to understand the scope of the problem domain, as per step 1 above. In-person interviews can be very effective in this initial stage to understand each user's background with respect to the problem domain, and their functional expectations and needs. The most appropriate interview approach is dependent on the familiarity of the visualisation researcher with the field in question.

Unstructured interviews are a useful initial tool to build familiarity with a problem domain. They work best when the interviewees are experts in the field who are acquainted with the successes and shortcomings of existing solutions, and are able to outline a range of potential use cases. This approach requires minimal knowledge of the field on the part of the visualisation researcher.

Semi-structured interviews have a similar exploratory aim as unstructured interviews, but are based around a short, fixed list of leading questions that help guide the conversation. These interviews are useful with users with less explicit existing requirements or familiarity with the field in question. A successful outcome is dependent on the visualisation researcher having conducted previous unstructured interviews, or being moderately acquainted with the field, in order to draft the initial list of leading questions.

Structured questionnaires are an appropriate choice when potential use cases or functions have been identified, and user responses are required to select or validate a subset thereof. This approach is well suited to collecting and collating responses where the potential sample of potential users is large. There is a risk that the focus of the questions can miss perspectives that would otherwise be unearthed in more personal interviews. It is therefore imperative that questionnaires should only be constructed after either of the previous interviews techniques have been applied, or if the visualisation researcher has a high degree of familiarity with the field in question.

Group sessions can interactively incorporate a plurality of perspectives, and the interaction between participants can lead to the discovery of new requirements or critiques that might otherwise be missed. These sessions can be used either in the initial discovery phase, or as part of feedback related to prototypes (discussed below). An enforced structure is required to manage the inherent noisiness of group interactions.

A potential problem with all the above approaches to scientific visualisations is the inherent difficulty of discussing the theoretical display of complex data in the absence of visual aids. While it is possible to use existing displays (where they exist) as departure points for discussion, additional aids will of necessity be required as the design study progresses. Iterative design through prototyping is powerful way of resolving this problem.

4.1.2.2 Iterative prototyping

The systematic review performed by Da Silva et al. [32] identifies the use of agile methodologies and iterative prototyping as useful in User Centred Design, an approach that places defined users at the core of the design process. Agile methodologies have also come to dominate the contemporary commercial software development landscape. They are a response to big, upfront designs that ultimately fail to deliver a product, or where that product fails to meet the needs of users.

These approaches are particularly useful where the full scope of required functionality is not known upfront and must be uncovered as part of an iterative design process. The development of prototypes that illustrate potential viable functionality is an important element in this process. The ability to interact with proposed solutions can elicit useful feedback early on in the process that would otherwise only be uncovered after final delivery. Prototypes also enable higher levels of dialogue about a problem by providing a common, visual context for discussion and experimentation.

The work of Jones, Marsden, et al. [46] complements this by suggesting an iterable, three-step approach to User Interaction Design. The initial step in each iteration involves gaining an understanding of potential user's capabilities, limitations and requirements. This understanding enables the development of a prototype interaction design that can be demonstrated, altered and discussed. The prototype can then be evaluated using a range of evaluation techniques. The outcome of this evaluation is ideally a list of strengths and weaknesses of the design, that can inform future iterations of the process. The evaluation may also support a more radical approach that discards much of what has been built in favour of a new line of design thinking that better addresses users' needs.

It is important to differentiate between prototyping and software development. Prototypes require only the minimum necessary functionality be implemented that is required to illustrate a particular function or interaction. They should use the simplest technology with the lowest barrier to execution, to support rapid changes driven by user feedback. There should also be no expectation that the prototype could easily be transformed into an eventual end-product - this prevents potential design ideas being rejected too early in the process due to potential implementation difficulties within a particular technology stack.

Paper prototypes are tactile, physical visualisation aids that have been shown to be helpful in exploring a range of possible solutions Snyder [79]. Their inherent simplicity and enforced abstraction are useful in rapidly generating and winnowing down lists of potential interface ideas. They can also be less intimidating to participants in the design process who are not familiar with software development. One drawback is that it can be difficult to map ideas to simple enough 2D paper models, particularly where the data or proposed visualisation is complex.

Static digital prototypes allow for rapid, reactive prototyping of changes or suggestions elicited in interviews. Static prototypes are the digital equivalent of paper prototypes, and are similar in nature to the design sketches used in the automotive industry. They are well suited as visual distillations of ideas generated during interviews and discussions, and can also be used for evaluation of those ideas during subsequent rounds of user interaction. Static prototypes are not well suited to illustrating the potential information channels afforded by interaction with a visualisation.

Interactive digital prototypes are much closer in capabilities to the potential final implementation, and directly address the difficulties static prototypes have in conveying interaction information channels. They can be more work to implement than other, simpler options, but with the right choice of tool set need not be. Exposing participants to a range of possible interactions can germinate new ideas for desired functionality. These can then be prototyped and evaluated, resulting in a virtuous circle of iterative design.

Participatory design

Participatory design includes users as integral parts of the design process, sourcing their input and evaluating their responses through the complete design cycle. This stands in contrast to the historic norm where users are consulted upfront and then presented with a completed product design (the waterfall model). Participatory design is complementary to agile methodologies in that it supports rapidly adapting to user ideas and feedback, as well as creating a sense of ownership and investment in users. This can often lead to a more thoughtful and thorough understanding and analysis of potential designs early in the process, and a greater sense of ownership of the solution by participants.

4.2 Previous work

The use of design studies for data visualisation is not currently widespread in the field of particle physics. Below we discuss four bodies of work that are either design studies in similar fields, or are developments of visualisation tools whose process adopts aspects of design theory.

Bärlund et al. [12] applied a user-centric design approach to improve the usability of electronic logbooks employed at CERN and ATLAS. Their approach utilised an initial fixed questionnaire, the results of which guided a participatory design process they describe as Contextual Inquiry (CI). Heuristic based user evaluation was conducted via usability testing in order to produce a proposal for improvement to be used in future logbook software development, including the CERN control centre *eLogbook* and ATLAS *ATLOG* electronic logbooks.

Chae et al. [25] conducted a visualisation-centric investigation into the use of colour and multi-scale histograms to support visual analysis of data from neutron scattering experiments at the Oak Ridge National Laboratory (ORNL) Spallation Neutron Source (SNS). Their approach utilised semi-structured user interviews to characterise the problem and define an initial design, that was then improved over several iterations in collaboration with potential users. In addition to a complete design, they also present several case studies illustrating the ways in which their design could be directly applied to user needs.

Weissenböck et al. [90] also involved users in their iterative, participatory design process to develop "An Interactive Tool for Exploring and Analysing Fiber Reinforced Polymers". Their final prototype was evaluated by users using a numerical scale questionnaire. This approach allowed the team to distinguish marked differences in perceived effectiveness between practitioners and scientists, for a specific technique.

The work of Tomberg et al. [85] presents a design study conducted with medical doctors. The aim of this study was to assist in "building representations and interpreting information" from episodic memories of encounters during their daily practice of medicine. Participatory design using paper models was successfully applied to create an initial prototype that was favourably received by users.

4.3   Expected outcomes

The primary outcome expected from this design process is a functioning interactive prototype that serves to demonstrate concrete examples of effective visualisation related to specific research questions. This final prototype is divorced from the implementation details that would be associated with integrating it into an existing tooling ecosystem. This choice emphasises the design-centric nature of this project, where the intent is to guide future development within an existing technology stack.

A secondary outcome is the documentation of both the process and experience of conducting participatory design with scientific experts where the visualisation researcher has minimal prior knowledge of the field. This may be useful to future design studies conducted in similar contexts.

An unexpected tertiary outcome was the use of the final prototype in the public outreach activities of CERN, which served as a broad, public validation of both the design process and the final prototype.

4.4   Approach

We synthesised common, successful aspects of previous design studies and techniques from literature to define the structure of our design study. We chose an approach that combines pure research oriented design study methods with iterative development derived from experience with commercial software development best practices. The broad outline of the plan is presented in Figure 4.1, which describes four distinct stages of work.

The basic approach consists of an initial foundational stage where relevant literature across physics, visualisation and design is reviewed. This is followed by a design stage comprising three iterations of the four phases of prototyping (analysis, design, implementation, and validation). A static **alpha** prototype is developed and evaluated by users in iteration one. This feedback is used to refine this into an interactive **beta** prototype in iteration two. This is again evaluated by users, and the feedback applied to finalise the visual and interaction design of the prototype in iteration three.

4.4.1 Foundational stage

The foundational stage provides an opportunity to familiarise the visualisation researcher with the primary concepts and basic theory underpinning the field under investigation, to be able to contextualise future discussions with interviewees. While it is possible to conduct interviews that rely solely on participant input, relying on interview subjects to explain basic concepts does not make optimal use of available interview time. It also becomes difficult to link ideas and draw conclusions without a contextual understanding of the field.

For this study it was deemed necessary to review and understand the theoretical basis of particle physics, and the software available and used within this community, which we summarise in Chapter 2. We then conduct a review of the field of visualisation in Chapter 3, where we explore the theory required to develop effective visualisations. We also review the range of visualisations and event displays currently available to scientists in the field.

4.4.2 Design stage

We chose to split the design stage into three iterations, each of which is further subdivided into four phases of design: analysis, design, implementation; and validation. These phases are further broken down into steps, each of which may address one or more of the levels of design and validation discussed in Section 4.1. The first two iterations follow all four phases. The final iteration focuses on incorporating the changes from previous validation phases in order to create a final prototype and thus only includes design and implementation phases. The key to Figure 4.1 shows which elements of the hierarchy of design are applied at each step within the process.

4.4.3 Iteration one

In the first iteration we identify a group of potential users of varying backgrounds. We then conduct initial interviews with them, the results of which are are transcribed, collated and aggregated to create a summarised list of common themes. This list is used to define the scope of the static **alpha** prototype.

The scope definition defines and constrains the expected functionality of the alpha prototype, in order to manage the required amount of work. It also focuses prototyping efforts on the minimum viable subset of functionality required to illustrate potential approaches and allow users to provide meaningful feedback.

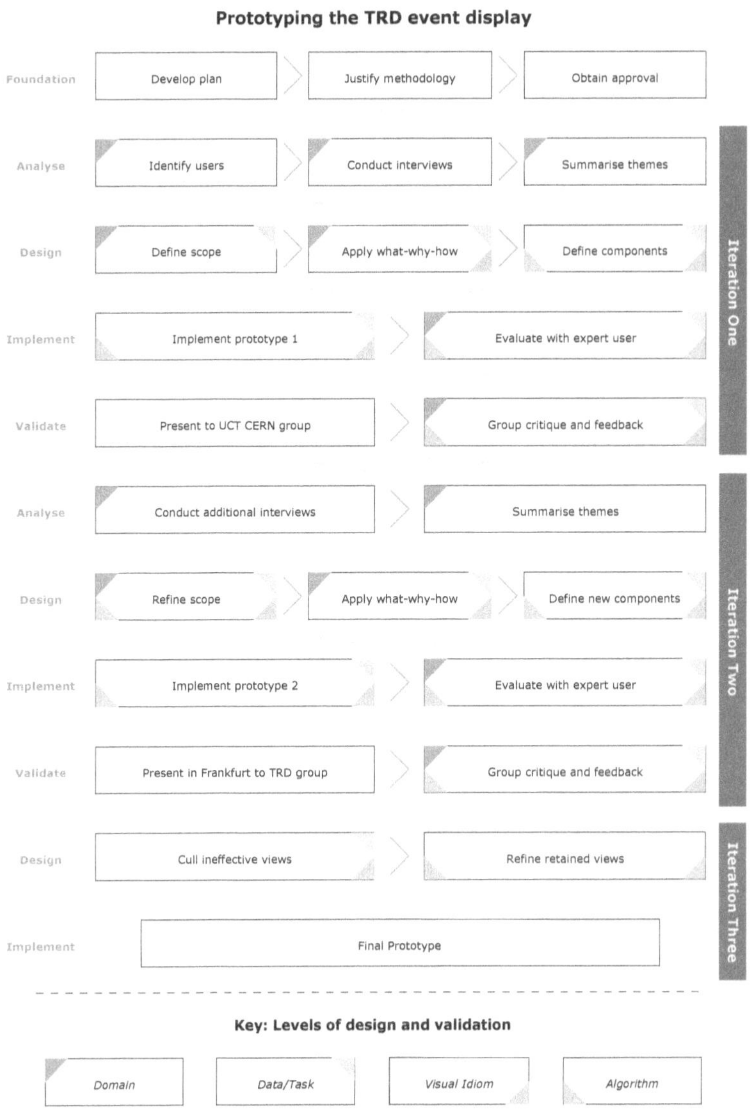

Prototyping the TRD event display

FIGURE 4.1: Overview of design study, organised in chronological order and categorised by design phase and validation level.

We then define a series of functional components based on the scope definition. The form and function of each component is gradually refined in a participatory design approach with the PDE. Reviews of, and feedback on, existing event displays are also used to guide the design of several components.

The resulting static alpha prototype is then evaluated in individual and group feedback sessions. The responses and critical feedback from these sessions is collated and analysed to again find common themes, which are used to inform iteration two.

4.4.3.1 Identifying users

We began by selecting a primary domain expert (PDE) to serve as a guide through the domain of particle physics, and provide high level direction and validation for the proposed solution. Dr. Tom Dietel is the current TRD software coordinator and has significant expertise as a particle physicist who works very closely with TRD data. He was able to explain unfamiliar physics concepts that arose during the process, and validate our understanding thereof, as well as assisting in navigating the existing software ecosystem.

It is infeasible to consider all potential audience profiles and expectations, due to time and capacity constraints on the part of both visualisation researcher and users. We therefore decided, together with the PDE, to consider only two potential audiences.

The primary audience are experimental physicists with a specific interest in the operation of the TRD and the resultant data. Their familiarity with the physical concepts underlying the operation of the TRD allows for more focused interviews with motivated participants. This audience would also be able to meaningfully critique the physical validity of visual representations, and suggest improvements.

The secondary audience is the general public with no knowledge of particle physics, with whom CERN would wish to conduct outreach and education activities. Visual appeal, clarity of information and a simple interface are more important than fidelity in such scenarios. This may conflict with the needs of the primary audience, but our expectation is that appropriate design choices can allow the initial single prototype to be easily customised for a different audience. We chose to focus on the primary audience for initial prototypes, with the secondary audience addressed once a robust and functional prototype was complete.

Potential users from the primary audience were identified by the PDE within the ALICE group, who were then directly approached to participate. The rights of the volunteers in terms of the approved ethical guidelines for the project were explained, in particular the rights to opt-out of the process and the right to anonymity.

4.4.3.2 Initial interviews

The initial interviews are the first opportunity to begin identifying the needs of users, and the visualisation tasks that directly assist research and analysis of data. We aim to build upon the basic domain knowledge acquired in the foundational stage, and better understand users' experiences and expectations in relation to visualisation. We also explore conceptual misunderstandings or areas of frequent misinterpretation of data, as well as uncover relevant knowledge gaps that can be bridged with appropriate visualisation aids.

An initial unstructured interview was conducted with the primary domain expert (PDE), the ALICE TRD software coordinator, to identify that a need for visualisation existed, and better understand the research area (to complement the field review above). The use of structured questionnaires was rejected in favour of interviews, due to the small number of identified participants and the variety of background and expertise between the interviewees.

Interviews were conducted primarily in-person at the University of Cape Town (UCT), with the use of video-conferencing for remote participants at the University of Heidelberg. The format consisted of a 30 minute core interview, with a 30 minute buffer for additional discussion if necessary; this was chosen to avoid question fatigue and keep the interview focussed.

An informal approach was adopted, using leading questions to prompt reflection, critical analysis, and brainstorming on the part of volunteers. Question and answer feedback loops were utilised, where answers to pre-defined questions prompted other specific, targeted questions. The common questions asked of every interviewee, and the expected outcomes, are reflected in Appendix A.

Interviews began with general background questions, to set the tone and understand the interviewee. We then established the field and domain knowledge of participants, as well as familiarity with event displays in general. Specific tasks were identified that related to the volunteers research interests, with emphasis placed on differentiating between personal and generally applicable requirements. This was important in the scope definition below, as common general requirements were prioritised over unique personal requirements. Volunteers were also prompted to identify aspects of existing displays that were problematic, as well as successful approaches that should be replicated, referencing existing work where appropriate.

4.4.3.3 Prototyping

For the initial static prototype, the option to use paper prototypes was rejected because sufficient examples exist of other event displays (see Section 3.5) to already establish a common visual language and use as a reference for discussions. A static web-browser based solution, optimised for high-resolution desktop screens, was instead chosen as the initial prototype development platform. This guaranteed portability and ease of setup when demonstrating to users. The web development stack directly supports rapid prototyping and encourages a focus on presentation rather than platform idiosyncrasies. A final important factor is that web development aligned well with this visualisation researcher's existing skill set, eliminating time that would otherwise need to be spent learning a different prototyping tool. The choice to use technologies with very little similarity to the core software stack used in the field[1] is a deliberate one that frees the potential designs from being constrained by the capabilities and expectations of that software.

We decided to construct the prototype as a series of independent components presented in a single interface. This would allow implementation of selection or rejection decisions from future reviews by simply adding, editing or removing the appropriate

---

[1]As discussed in Chapter 2.3.

components. This choice also aligns with existing approaches[2] and specific user requests from the interview process. This approach standardises user eye movement between identically sized, regularly spaced components. Finally a component approach allows for parallel prototyping, as changes to a single component can be implemented in isolation to others. We attempt both low and high fidelity spatial representations to determine effectiveness and potential value. The component approach also supports addressing the needs of the public audience at a later stage by modifying existing components or adding new components as required.

Interaction for this initial prototype was limited to ensure less upfront implementation was required. Users would be able to select from a predefined representative set of events, and the individual components would update accordingly, but all other aspects of the display would remain static. This choice is in keeping with the principle of using the simplest visual information channels first, before considering more complex but powerful options.

4.4.3.4 Validation

The static alpha prototype was presented in a group setting to an audience of subject matter experts and potential users at an ALICE group meeting in Cape Town, South Africa. This presentation demonstrated the capabilities of each prototype component, individually and in relation to other components. A question-and-answer session was conducted after the presentation to elicit feedback from the audience.

Specific feedback was elicited on the effectiveness and appropriateness of each potential component, as well as general feedback on the approach and design choices evident in the alpha prototype. Follow up questions to the audience, based on this feedback, were then used to expand upon and clarify the critical feedback on these components.

A second interview with the primary domain expert (who was initially interviewed) was conducted after the group session. The aim was to evaluate the static prototype in greater detail than in the group setting, as well as clarify any points raised by the group that depended on specific domain knowledge. The PDE was also asked to provide feedback on whether, and to what extent, points raised in their first interview had been addressed, and whether this had raised any potential further requirements based on the demonstrated prototype functionality.

This prototype was also presented to Assoc. Prof. Michelle Kuttel, an experienced visualisation designer, who provided valuable critique on the visual design choices (colours, textures, styles, layout), which was then incorporated into iteration two.

4.4.4 Iteration two

In the second iteration we again follow the four phases of design illustrated in Figure 4.1. We begin by conducting another round of interviews with a new set of potential users, as well as attending several presentations on the development of an AliEve replacement in O^2 . The interview responses, presentation content and static prototype feedback (both direct and indirect[3]) are summarised to extract common themes and revise the overall scope. Identified problems with clarity of information representation or ease of interaction are specifically addressed. Interactivity is incorporated by

---

[2] As demonstrated by AliEve (Figure 3.5)
[3] Observed responses during group presentation that were not explicitly vocalised.

allowing users to zoom and filter on data of interest, as well as using animations and transitions to aid understanding.

We then summarise the common themes and incorporate the experience gained in building the initial static alpha prototype, creating a list of concrete changes to existing components, and additional new components to implement. These changes are applied to the existing static **alpha** prototype in a collaborative design process with the PDE to create an interactive, dynamic **beta** prototype. This beta prototype is again critically evaluated and validated in both individual and group sessions, and the consolidated outcomes used to finalise the design.

4.4.4.1 Additional interviews

The second round of interviews involved a new set of participants again identified by the PDE, all of whom represented the primary audience for the display. The rights of the volunteers in terms of the approved ethical guidelines for the project were explained, in particular the rights to opt-out of the process and the right to anonymity.

All interviews were conducted in-person either in Geneva at CERN or in Cape Town at the University of Cape Town. A similar format to the initial interviews was adopted using the leading structured questions from Appendix A. The alpha prototype from iteration one was also demonstrated to participants to prompt further discussion and idea generation. User feedback was used to select a subset of components that were judged to be useful and effective, and the critical responses were interpreted to improve the functionality and appearance of these components.

4.4.4.2 Validation

The developed interactive beta prototype was presented to an international audience of subject matter experts at the *TRD Comprehensive Status and Planning Meeting* in Frankfurt, Germany. The audience represented varying degrees of familiarity with, and expectations of, event displays, and the majority had not previously been exposed to our design process. As such they were able to provide valuable feedback from the perspective of new users of the visualisation, as well as provide novel critical analysis not previously encountered.

We conducted a hands-on case study at the meeting with a group of physicists with a specific interest in TRD data. This session served both as an evaluation of the beta prototype, and a source of very technical corrections to some of the implemented functionality, as discussed in Section 5.3.4.

Upon returning to South Africa, the beta prototype was presented at a combined ALICE and ATLAS group meeting in UCT, where feedback from expert physicists outside ALICE was obtained for the first time. One final interview was then conducted with the PDE to discuss and contextualise all the feedback received, as well as discuss the successes and failures of the beta prototype given the initial remit.

4.4.5 Iteration three

The third iteration is a final refinement of the work produced in the previous iterations. We synthesise all prior feedback to identify a final set of components to retain, document the desired functionality represented by each component, and then update the **beta** prototype to reflect these changes. The **final** prototype is then demonstrated and used by several physicists on their own data and these case studies are briefly summarised.

During this process the opportunity arose to include the final prototype as part of a large public outreach event at CERN Open Days [22], and elicit feedback from the original secondary audience of this project. This required changes to the final prototype in addition to those identified as part of previous feedback. The details of this successful demonstration, and the impact it had on the final design are also documented as part of this iteration.

5 Results: fruits of our labour

This chapter outlines the results of our design study to create an event display for use within the TRD group in ALICE at CERN. We present the details and results of each iteration of design, using the methodology and evaluative research based methods outlined in Chapter 4.

We begin with iteration one and present the common themes summarised from user interviews, and then outline the resulting scope definition of the **alpha** prototype. We then provide detailed descriptions of the chosen components, and discuss implementation specific considerations that we encountered. We finally summarise the responses from validation of the alpha prototype, and document our personal reflections on the process.

For iteration two we present the common themes summarised from additional interviews conducted after demonstrations of the alpha prototype. We then detail the revised scope for **beta** prototype, and describe the changes to existing components and functionality of new components, as well as associated implementation considerations. The feedback from presentations of the beta prototype are then summarised, and we discuss personal reflections on the process.

We conclude with iteration three where we discuss the presentation of the **final** prototype at CERN Open Days and the changes to the design that this necessitated. The final design is presented and discussed in detail, as are the associated implementation details. We finally summarise all the case study demonstrations that were conducted to evaluate the prototype and related outcomes.

5.1 Alpha prototype: Iteration one

5.1.1 Initial interview themes

We conducted interviews with four volunteers, referred to hereafter as PDE [36], P2 [59], P3 [60], and P4 [62]. The volunteers' had a range of experience with the TRD, its data and visualisation thereof: one novice with no previous experience; one intermediate with limited analysis exposure; and two experts involved in detailed analysis and event reconstruction. All volunteers had completed at least a 4-year degree in Physics, and the range of experiences was sufficient to cover a large set of potential expectations and tasks. Feedback has been grouped by similar themes expressed by one or more participants, identified by the corresponding abbreviation in parentheses.

Tasks

The primary motivation behind a new event display is to support the move to the new O^2 framework[1]. A TRD specific display would help debug and guide development of both simulation and reconstruction software, as well as assisting with detector calibration and error detection (PDE, P3, P4).

There also exists a need to reconcile TRD specific data with the reconstructed data from the rest of the ALICE detectors. In particular, it was deemed important to be able to illustrate the relationship between global tracks reconstructed by ALICE and tracks reconstructed by the TRD using tracklets (PDE, P2, P3, P4). The ability to interactively select and compare multiple tracks would further assist this task (PDE) and would improve on the manually constructed ROOT histograms that are the current primary visualisation used (PDE, P2).

Participants felt that existing software did not effectively illustrate the detailed operation of the TRD, and specific effects unique to it (PDE, P2, P4); the representation of tracklets in particular was highlighted.

Finally, a common theme was the use of a simple event display as an introduction to the experiment in general, and the TRD in particular (PDE, P2, P3). Participants felt that visualisation would help new users better appreciate the geometries involved, and the relationship to the data that they would be required to analyse. This use could also be extended to outreach activities to non-scientists.

Data

An important consideration for the expert users was the need to support data from both simulations of Run 3 as well as recorded data from detectors to allow cross-validation between the sources (PDE, P4). This capability should include the ability to bring together disparate data not normally co-located (PDE).

Participants made clear that in most cases they are interested in a limited subset of the data (PDE, P4). A full heavy-ion Pb-Pb collision contains too much raw data, and hence users wish to visualise post-cut[2] data.

Platform

Portability of the final solution was important to all participants (PDE, P2, P3, P4). The responses suggested the ability to run on many platforms could: broaden the potential audience; reduce barriers to entry, increasing use within the field; and provide the ability to share visualisations with colleagues to easily illustrate areas of interest.

Users reported that they often do not have time or resources to focus exclusively on tool development (P4). Existing event displays are difficult to setup and run, and in particular AliEve[3] is not included in their normal workflow due to the time investment, complexity and manual setup required (P2, P4). A display focused on ease of use, simple setup with only minor configuration required, and able to run in a low resource environment was deemed valuable (P4).

---

[1] Discussed in Section 2.3.5.

[2] Particle physics terminology referring to data after a series of filters have been applied to exclude background noise.

[3] Discussed in Section 2.3.3.

Display

It was suggested that several linked views of the same dataset aid understanding (PDE). This would allow a user to visually switch between projections without interaction, and compare different data points in a standard way. This requirement is echoed in the work of Drevermann, Kuhn, and Nilsson [38] who also identified the value of several synchronous views of the same data, each showing different projections and magnifications.

Participants expressed a need to customise aspects of the display, toggling the visibility of individual elements and geometry to better manage and filter the complexity of data displayed (PDE, P2, P4).

3D-visuals were highlighted as a useful approach to providing an overview of the experiment and data. 3D-visuals were criticised for the difficulty involved in both judging distances between points and lines, and understanding details of the data (PDE, P4). They were deemed less useful for investigating specific areas of interest, as there is generally no automatic way to focus on a single event and the user must continually cycle between different views.

2D projections were regarded as a useful supplement to 3D that are easier to understand, and well suited to investigating behavioural details (PDE, P2, P4). Predetermined, fixed, 2D-projections would be helpful to standardise orientation and eliminate the view-cycling required in 3D. The ability to display textual information relevant to the displayed data was also requested (PDE, P4).

Interaction

Participants reported that ROOT based 3D-displays were difficult to rotate and zoom to desired views of the data (P4). Further discussion suggested using multiple fixed zooms with 2D projections, which would allow both overview and detail to be presented in the same visual space, removing the need to manually zoom in and out, resulting in an easier to use interface.

It was also suggested that the previously mentioned linked views be used to show data in parallel, thereby negating the need to switch back and forth between related datasets (PDE). Animated transitions could help maintain user orientation as the displayed data changed in response to selection changes.

Novel functionality

During the interviews, participants were prompted to brainstorm novel ideas or functionality not available elsewhere. The most common suggestion was the ability to link the event display via an interface to other software (PDE, P2, P4). Such a link would allow the event display selection to be externally driven or enable multiple users to collaboratively discuss a single shared display. This interface could also be used to link the displayed data to the results of an external physics analysis, or simply display data directly from the CERN grid storage rather than having to download it manually first.

A more technically challenging request was to illustrate the complete process by which raw ADC data[4] is reconstructed into tracklets, which in turn are used to reconstruct

---

[4] A whole number indicating the quantity of charge deposited in a chamber, as generated by the Analogue to Digital Conversion process.

tracks (PDE, P4). This would allow users to investigate how the ADC values recorded over time-bins match the reconstructed tracklets. Users could also check that tracklets are attached to the appropriate tracks (P4), which is important in validating the reconstruction algorithms for Run 3.

A participant suggested that animating the progression of a particle collision over time would be useful in assisting novices understand: the sequence of particles produced; the related interactions with detector elements; and the corresponding signals measured (P3).

More generally, there was agreement amongst expert users that a design centric exercise would be useful to gather opinions and direct future development, as such an endeavour has not previously been carried out in the context of the TRD (PDE, P4).

Visual queries

When prompted, users suggested a number of specific data-related questions that they would want to investigate using an event display:

- Why is a track far from its matched tracklets?

- Why didn't a track match nearby tracklets?

- Why are so many tracklets not linked to tracks?

- Is the overall charge deposition sensible?

- Are there any hot or cold spots (areas with high or low charge deposition)?

5.1.2   Scope definition

A common theme amongst users was a need to support multiple types of data. We decided to limit the initial alpha prototype to displaying reconstructed data only, and consider other options in subsequent prototypes. The primary motivation is that reconstructed data is easily available (as it is already used for physics analysis) and is highly processed, so less time is spent on data preparation and cleaning. This also adheres to the visual principle of *overview first, then zoom and filter*, by postponing visualisation of detailed data until the high-level representations are resolved. We enable this by utilising the data stored in AliESD[5] files, which store the processed and reconstructed data across all ALICE detectors.

The majority of users' responses expressed an explicit interest in tracks and tracklets, as tracklets are TRD specific data used to reconstruct the tracks of particles created in a collision. We therefore chose to focus on visualisations of these elements in order to aid contextualisation of available data, by supporting an understanding of the link between these two artefacts. This approach links two different levels of reconstruction and can also be used to understand errors in reconstruction as requested by users. For simplicity we consider only approximate tracklet location, and defer complexities of detailed representation to the subsequent iterations, as it is not yet clear if is required.

We additionally choose to focus on Run 2 data, as the data content and format is fixed and established methods exist to extract this data. Real data available from experimentation is available, that would otherwise have to be generated via simulation for Run 3. One important consequence is that we do not consider the consequences of the

---

[5]See Section 2.2.4

(A) Tree navigation component.

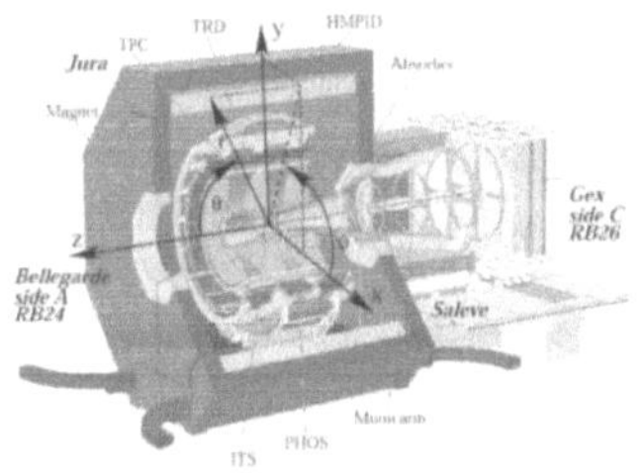

(B) ALICE with coordinate system [14].

FIGURE 5.1: Components to: A) navigate through events; and B) illustrate the ALICE coordinate system.

Run 3 continuous data taking proposals, as these were still in flux at the start of the project.

Many examples of generic event displays that support ALICE exist[6], hence we have chosen to focus our visual prototypes on the specific contributions of the TRD, based primarily on the documented interests of our user group. We further limit the potential solution space by effectively ignoring other detectors, and specifying that display component prototypes all use TRD-specific fixed data.

The initial components consider 2D projections only, adhering to the *no unnecessary 3D* visualisation principle. Such projections provide a simple starting point that users are familiar with, and discussions using them can uncover requirements for more complex representations. We also consider variations on existing projections, as well as novel new proposals.

5.1.3 Component definition

The following sections detail the components utilised in the alpha prototype for iteration one. They are based on the user interviews we conducted, guided by the scope definition above.

5.1.3.1 Navigation components

These components allow the user to navigate the event display. The choice to support multiple collision events requires an interaction component that facilitates selection between events, and sub-selection of a specific track within an event. We have chosen the standard idiom of a tree navigation component, as illustrated in Figure 5.1A, which is familiar to most users.

The list of available events is listed at the top level, with the available tracks listed as selectable child nodes. Tracklets matched to tracks are listed as child nodes of the corresponding track, and tracklets not matched to any tracks are grouped together under a single child of the event named *Tracklets*. Events are labelled with their unique ID from the input data, while tracks are labelled with their ID[7] as well as the stack and sector of the module they pass through.

---

[6] Some examples are discussed in Section 3.5.

[7] Unique within an event, but never globally unique.

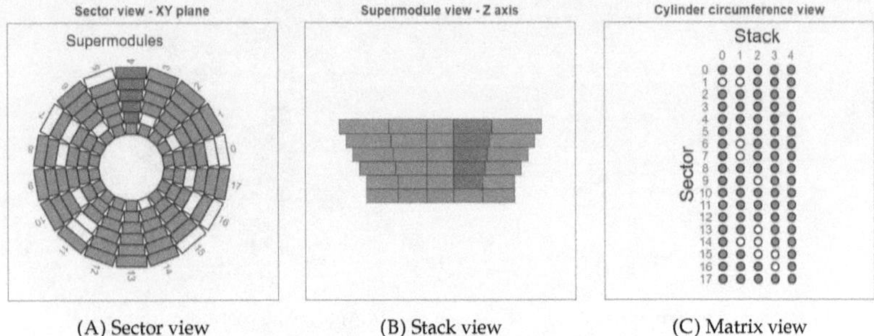

FIGURE 5.2: Three different representations of the TRD, with modules shaded such that: no shading indicates an absence of tracklets; gray indicates at least one tracklet was recorded; and red indicates a tracklet associated with the selected track was recorded.

Each node in the tree can be independently expanded or collapsed, allowing the level of visual clutter in large datasets to be dynamically managed by the user. This tree is linked to other components in the display such that selections in the tree cause other components to dynamically update, displaying progressively increasing levels of detail.

Figure 5.1B is a static image of ALICE, adapted from Betev and Chochula [14], showing a cut-out view of the detector internals. This is potentially useful as a way to conceptually link the various 2D projection component views to the corresponding 3D spatial layout. This image also includes a visual representation of the standard ALICE coordinate system (described in Section 2.2.2) as it relates to the various sub-detectors, including the TRD in particular. This is important context when interpreting the labels and scales on the various components that follow.

5.1.3.2 Tracklet components

Tracklets are the unique data contribution of the TRD, and three components were identified that could be effective in visualising this data. These views of the data prioritise spatial representations that clearly indicate which detector modules registered *at least one* tracklet. This is useful for analysing the distribution of tracklets within the detector, as well as supporting visual pattern matching to locate anomalies. The relative size of detector modules have been distorted in these views to increase visibility.

Figure 5.2A is the familiar sector view, looking down the beam pipe. Each block represents a horizontal layer of five modules; six layers are stacked within a single sector. The radial size of each module is greatly exaggerated to make the most of available space, and this is repeated for all 18 sectors radially. Dark gray highlighting is used to indicate that at least one tracklet was reconstructed by the corresponding module. If a specific track is selected in the navigation tree, any associated tracklets[8] are highlighted in red. The chosen colours differ in both hue and luminosity, ensuring they are easily differentiable.

---

[8]Discussed in Section 2.2.4.

Figure 5.2B is the corresponding stack view, with the same visual encoding applied. We have projected all 18 supermodules (and associated tracks and tracklets), via rotation about the z-axis, on to a single representative supermodule[9] to better utilise available space. In this view, a dark gray rectangle indicates that the corresponding module in at least one supermodule recorded at least one tracklet, and if a track is selected a red highlight indicates the presence of a matched tracklet. This choice is based on feedback that indicated most tracks (and hence matching tracklets) of interest pass entirely through a single supermodule.

Figure 5.2C illustrates an alternate approach to encoding tracklet positioning. In this view we have mapped the combinations of stack and sector into a rectangular grid. As per the previous colour encoding, a shaded circle then indicates that at least one layer within the highlighted stack/sector pair recorded a tracklet, and a red highlight indicates the tracklet was matched to the selected track.

Across all three component views, the primary supported visual tasks corresponding to identified user needs were:

- Identifying the modules in a which a measurable signal was deposited by a transiting particle.

- Understanding the spatial distribution of all recorded tracklets.

- Locating potentially problematic modules that never record tracklets.

- Understanding the relationship between tracks and matched tracklets.

- Identifying modules where no tracklet was recorded, particularly for the selected track.

- Verifying tracks and matching tracklets are spatially located in close proximity.

5.1.3.3 Track components

As described in Section 2.2 the track followed by a particle through the detector is important in determining the momentum and charge of that particle. Figure 5.3 illustrates three potential designs to visualise the magnitude and direction of the curvature of a track.

Figure 5.3A is a sector view of the TRD, similar to Figure 5.2A but with an abstracted cylindrical cross-section (shaded light blue) that is scaled to accurately represent the relative size of the TRD within ALICE. The path of a single particle is shown as a bright red visually distinct line arcing from the collision point through the detector. This orthogonal projection makes clear the effect of the magnetic field on the charged particle, inducing a helical trajectory. A distance scale in centimetres appears on both axes to help quantify observed features.

Figure 5.3B is a stack view of the track in the same visual style. It is similar to Figure 5.2B but uses a slightly different mapping from 3D to 2D space. In this mapping each supermodule is individually rotated about the z-axis until its centre lies on the y-axis. All tracks passing through a supermodule are then rotated about the z-axis with the corresponding rotation before being orthogonally projected onto the zy-plane. This

---

[9]Previous work has generally used two as in Figures 3.2, 3.5 and 3.6.

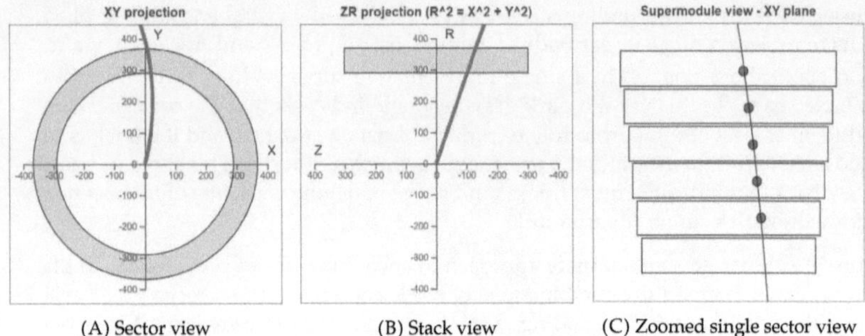

(A) Sector view   (B) Stack view   (C) Zoomed single sector view

FIGURE 5.3: Three different representations of the path of a particle passing through the detector.

is labelled the *Z-R projection* in the view as it is enables comparison of the radial behaviour of all tracks relative to the z-axis. This view of the helical trajectory perpendicular to the direction of travel is a sine curve with a period related to the momentum of the particle – the oscillating shape we expect is only visible in low-momentum tracks.

Figure 5.3C is a zoomed-in version of Figure 5.2A that shows both the track of the interacting particle and the TRD tracklets that were matched to that track. This view allows a user to visually determine what fraction of the six potential tracklets were attached to a reconstructed track, and whether those chosen appear feasible.

Across all three component views, the primary supported visual tasks corresponding to identified user needs were:

- Understanding the shape (curvature and direction) of particle tracks.

- Querying the relationship between tracks and tracklets.

- Comparing multiple tracks overlaid on each other[10].

- Verifying that selected tracks do intersect the TRD.

- Verifying that tracks have the expected minimum number of matched tracklets.

- Identifying cases where the apparent match between track and tracklets appear poor, warranting further investigation.

5.1.3.4 Novel 2D projections

The group of components illustrated in Figure 5.4 are all attempts at novel orthogonal projections of a single particle's track through the detector, from a helix in 3D[11] space into 2D spaces that are not commonly utilised. They do not directly address a specific user need, but are instead intended to expose users to potential modes of representation they may not previously have considered. Successful representations would illustrate features that are poorly discernible or hidden in other representations. This in turn could inspire ideas or requirements for future phases and prototypes.

---

[10]This is not explicitly illustrated, but was discussed with users during validation.

[11]As discussed in Section 2.1.6.2.

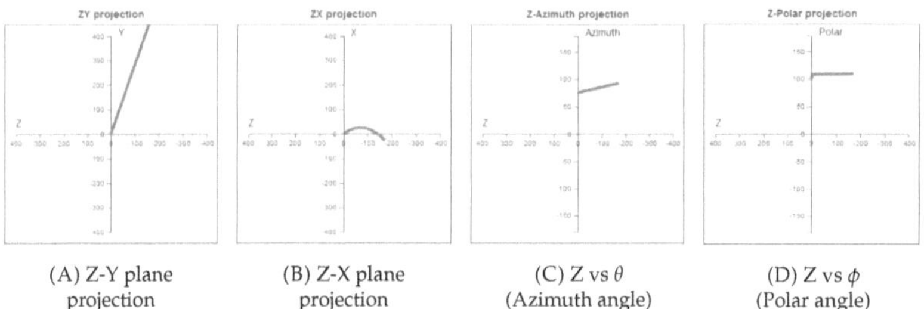

(A) Z-Y plane projection	(B) Z-X plane projection	(C) Z vs θ (Azimuth angle)	(D) Z vs ϕ (Polar angle)

FIGURE 5.4: Four attempts at novel 2D orthogonal projections of a particle track.

The first pair of components project the track onto planes that lie on the z-axis. Figure 5.4A is an orthogonal projection of the particle track onto the zy plane, without the rotation transformation from Figure 5.3B. The corresponding orthogonal projection onto the zx plane is illustrated in Figure 5.4B.

The second pair of components graph the polar coordinates (θ and ϕ in Figure 5.1B) of the track in 2D. Figure 5.4C is a plot of the azimuth angle θ against z and Figure 5.4D plots the polar angle ϕ against z.

5.1.4 Implementation

Given the previous discussions, we chose to prototype from scratch rather than building on top of existing code. The prototype was packaged as an **HTML** file, an internet standard that runs everywhere in a browser, requires no installation, and has a lightweight footprint.

Individual components, as defined above, were implemented as dynamic **SVG** images. SVG is a vector-based, scalable graphic format that ensures crisp images with minimal processing overhead and good performance even in low resource environments.

These images were dynamically built using **JavaScript**, the ubiquitous scripting language of the web, and **D3.js** [15], a powerful data visualization library that supports generating highly customised, interactive SVG. **JQuery** [48], a standard utility library, and **JQueryUI** [47], a library of web user interface components, were also used for layout, navigation and asynchronous data loading.

JSON is a web standard for data storage, and was used as the common format for both the reconstructed data (tracks and tracklets) as well as the detector geometry definition. We chose to use JSON, this rather than the existing ROOT binary object format, as JSON is a text based format that is easy to read and generate, and its use has been standardised across the internet.

We defined a standard, intermediate JSON schema for ALICE event collision data in Appendix C that is abstracted from a particular framework. This allows development to be decoupled from changes in the data source, especially relevant given the data specifications of O^2 were still in flux, as discussed in Section 2.3.5. This schema allows data from ROOT, O^2 or simulations to be displayed in the same event display – all that is required is a conversion utility to translate from the desired source format.

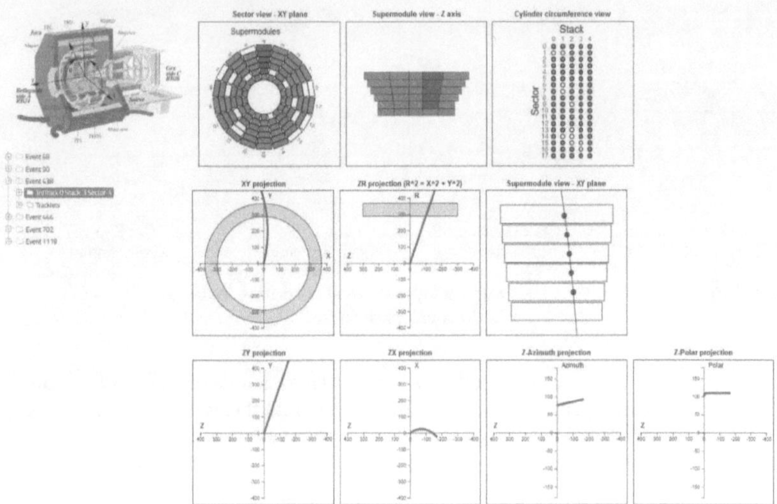

FIGURE 5.5: Screen capture of entire iteration one alpha prototype as
presented at ALICE group meeting.

We further implemented such a conversion from the existing AliROOT AliESD file
format[12], the details of which are discussed in Appendix C.

This JSON schema definition is an important artefact that future work can utilise and
extend. It provides a common target format that is easily parsed, enabling simple data
sharing without the need for specialised tool chains. The common format also allows
future displays to focus on visualisation rather than needing to understand source-
specific complexities. Finally it supports the goal of familiarising novices with TRD
data by being human readable, and the hierarchical nature of the format allows data
elements that are not relevant to a user to be visually hidden in modern editors.

There were several additional complexities that had to be understood and overcome
during development, the most salient of which are documented in Appendix D. The
complete iteration one prototype appears in Figure 5.5[13] and is available online at
`https://datacartographer.com/alice-trd-event-display/v1/` [69].

5.1.5 Feedback summary

The alpha prototype was presented to a group of subject matter experts and potential
users at an ALICE group meeting in Cape Town, South Africa. A second interview
with the primary domain expert (who was initially interviewed) was conducted after
the group session. The alpha prototype was also presented to visualisation expert
Assoc. Prof. Michelle Kuttel.

The responses and critical feedback from these presentations were collated and analysed to again find common themes, which were ranked according to user value and
required effort. The most relevant are presented below, and inform the scope, design

---

[12]See Section 2.2.4

[13]A larger, full-page version appears in Figure D.5.

and implementation of iteration two. They are grouped according to whether they relate to: the visual display of information; the manner in which the user navigates the interface; or modifying the functionality of the prototype.

Display

Many responses indicated that there was redundancy between the track (Fig 5.3) and tracklet (Fig 5.2) views. The use of the same colour for both tracks and tracklets also led to confusion. Users suggested overlaying tracks and tracklets in a single view, and colour coding them differently. Users also requested that multiple tracks be shown alongside the selected track, to better facilitate comparison.

The distortion of size and aspect ratio in tracklet views was also deemed to be unhelpful, as this distorts the relative size and shape of tracks. This is problematic as the ability to view the curvature of the entire track is important to estimate the momentum of a track. The tracklet zoom view was considered effective, but users requested the detector sub-structure and the interaction point be visible in order to better understand the overall event. One suggested solution was to use a set of fixed zoom views that allow both overview and detail to be visible at the same time, while also maintaining size and aspect ratio.

Tracklets were identified as inaccurately positioned as point data at the centre of chambers. Expert users highlighted that tracklets are actually 2D parallelograms running the length of an individual pad, and are tilted both in the x-y plane to reflect the path of the incident particle, as well as being tilted by $2°$ in the Z-direction relative to the chamber center.

Users questioned the simplified detector geometry used, with the five chambers per layer represented by five equal sized rectangles. The geometry sourced from [5] specifies that the chambers in stack two are all the same length (z-direction), while the dimensions of layers zero and one are identical. It is also necessary to account for the dimensions of the supporting spaceframe and related infrastructure. A further complication is that the detectors and spaceframe can both deform after installation due to the weights involved. All this necessitates that the actual geometry used should ideally be obtained from the ALICE alignment data, which is updated approximately once per a year.

The additional 2D projections that were attempted (Fig 5.4) were not found to be useful, and the cylinder matrix view (Fig 5.2C) was not effective at correlating 2D projections with 3D space. It was suggested that all be removed and the remaining components enlarged to make better use of available space.

Navigation

The ALICE image overview (Fig 5.1B) was considered unclear and not helpful by users, given that the static 3D image was too low resolution to be useful and takes up available space better used for interactive elements.

The tree selection component (Fig 5.1A) for selection was deemed to work well, but the ability to select individual tracklets seemed unnecessary and added visual clutter to the display.

Novel functionality

The conversion task from ROOT files to JSON[14] was deemed very useful. Users requested the additional ability to provide an input file with numeric identifiers (run, event, track), and have all matching data automatically exported by the same task.

A common, if expected, request was the ability to view the raw data associated with tracklets[15]. One suggestion was the ability to click on a track to view both the matching tracklets and the corresponding underlying raw data.

While some users indicated that a 3D representation would be useful, questions were raised by others as to how to synchronise the 2D projections of a displayed event with the corresponding 3D contextual display, particularly in terms of camera location, rotation and zoom.

Users also requested additional information on the selected track be displayed in a text-box, with specific mention of the collision energy, particle momentum, and identity, as reconstructed from other detectors in ALICE.

5.1.6   Reflections on the process

Although the end result of the first iteration is essentially a series of very simple pictures, it required a long and winding journey to arrive at those pictures. The very technical nature of the domain required a large allocation of time spent working through the details of software, data, and theory to be able to ask the right questions in the interviews! It is our personal hope that the output of this research will enable a much larger audience to achieve the same understanding without an equivalent level of required effort.

We found it useful to progressively manage domain complexity by intentionally ignoring certain unknowns to get a best first-order understanding, which could then be expanded through discussion and feedback. Understanding the common coordinate systems used in ALICE was particularly helpful in expressing and understanding particular ideas more concisely.

The use of highly simplified components that illustrated distinct ideas or information was useful during prototyping. Users were able to quickly grasp what they were seeing, and this understanding often prompted requests for more complex, detailed functionality. This approach enabled us to verify that the correct data was being presented in a meaningful way.

We found that robust interactions in group settings were an effective way to elicit focused feedback on specific aspects of the alpha prototype. The collaborative discussions involving multiple voices greatly assisted in understanding the various, often subtle, physical effects, as well as the concerns of potential users.

---

[14]See Appendix C.

[15]Within the TRD group this is often referred to as digit or ADC data.

5.2   Beta prototype: Iteration two

5.2.1   Secondary interview themes

The second round of interviews were conducted with the PDE [35] and five expert users (P5 [61], P6 [63], P7 [65], P8 [66], P9 [67]), all of whom had at least a PhD level qualification in particle physics or were in the process of acquiring one. One of the participants (P6) is directly involved with the development of an AliEve replacement in O^2 . Their experience offered unique and valuable insight into the challenges and opportunities encountered developing a similar tool focused on hardware accelerated 3D display of data from every detector within ALICE. We attended their presentation to the ALICE community on progress with this project [64], and the salient points from that were merged with the interview responses.

These responses are grouped into three broad themes below: expanding the audience for event displays amongst users and developers; comparisons with similar, existing work in the field; and opportunities for future development or functionality.

Expanding the audience

Some of the second-round participants were specialists involved in very technical aspects of the TRD, such as electronics developments or upgrades (P7, P8), and as such were not necessarily very familiar with the use of the TRD in physics analyses. The alpha prototype nevertheless seemed valuable to them in contextualising their work, and providing a simple entry point to gaining an understanding of the higher levels of TRD operation. They also highlighted opportunities for visualisation of data below even tracklet and raw data level, functionality which is not commonly considered when designing a display only from a reconstruction and analysis standpoint.

The International Particle Physics Outreach Group hosts *Master Classes* [45] that introduce novices to the organisation, its experiments, and their data. Users suggested using the developed prototype in these sessions, as it appeared to provide a simplified environment for novices (PDE, P6). The self-contained, web-based deployment enables simple setup on multiple computers, and the interactive interface would allow attendees to visually identify areas of interest (such as tracks, tracklets and the interaction point) with ease.

Users also commented that the event display alpha prototype, as presented to them, could be used to debug the TRD, generate images for publication, or for knowledge sharing (PDE, P5, P6, P9). This highlights the value of pre-packaged, easy to use visualisations to technical users whose primary research focus does not include visual design and development.

Comparisons with related work

Technical debt, developer resource availability and the use of perceived older technology such as C++ were highlighted as obstacles to updating and extending existing event displays (P5, P6), which makes it difficult to attract students who are interested in more recent languages and technologies.

Web-based technologies such as **threejs** [84] have been considered during development of an AliEve replacement as an attractive alternative with a desirable supporting technology stack (P6). At present there is little appetite for a general move to such a platform, given the familiarity and experience that resides within the ROOT ecosystem. It was also stated that the assumption that browser-based interfaces require more memory and processing power are also an obstacle, despite recent versions of Root having web-based capabilities.

The development of the event display for O^2 highlighted the need to balance the display of all available data against available resources, given the sheer volumes generated by ALICE in even a single collision. Significant effort is required to optimise and utilise hardware acceleration to support dense data representations. Our decision to focus on small subsets of data within single events mitigated this issue to a large degree, allowing us to focus primarily on visual representations.

A related issue experienced in O^2 was the difficulty developing for Run 3 in the absence of sample data (P5, P6, P7). This was exacerbated by the in-flux status of data formats, and the tight coupling to the event display required for efficiency reasons. JSON was proposed as an eventual ultimate store of visualisation data stored in a general format that is usable by multiple detectors and experiments [64]. This validated our decision to use JSON as an intermediate format for prototyping.

Future opportunities

Run 3 data is planned to be continuously recorded, with all data recorded within an interval of 256 heartbeats[16] (22 ms) aggregated into a single time-frame for processing and storage [72]. A time-frame based display would be helpful in allowing users to interactively navigate and inspect the data in a single time-frame. One suggestion was the use of a slider to select a subset of the data, and animation to convey the passage of time (P6). Visualisation would also be helpful in helping users translate their existing data knowledge to the new O^2 approach.

Controlling the event display via simple scripts, connecting to remote data sources, and displaying synchronised views of the data in parallel across multiple screens or machines were additional functionality requested by users (PDE, P5, P6). These capabilities would enable visualisation to be provided as a standardised, centralised service, rather than requiring custom knowledge and installation on the part of users.

Users also expressed an interest in displays for live status and data quality monitoring during data acquisition (P5, P9). Such functionality would update the existing visual monitoring of TRD status attributes during LHC operation (such as temperature, voltage, and drift velocity), representing an improvement on the existing display illustrated in Figure 3.4. Reference was also made to SMMon (*SuperModule Monitor*) (PDE, P8) an existing, unix, text-based monitor for individual supermodules within the TRD. A simple prototype web-interface that addresses this need is discussed in Appendix D.

5.2.2 Revised scope

Interviewee responses that suggested the addition of new visualisations or data sources were incorporated as new components within the beta prototype, to be evaluated alongside the retained existing components. The visual style across components was also standardised to create a cohesive aesthetic that neither clashed with nor distracted from the primary information visualisation. Component sizes and layouts

---

[16]A heartbeat in this context refers to the time taken for a single orbit of the LHC loop.

were optimised for large, high-resolution screens, to better utilise all available screen space and maximise information density.

As per user feedback, the tree navigation component was retained (Figure 5.1A) but the ALICE overview image was removed (Figure 5.1B). The novel 2D projections (Figure 5.4) were all removed for being ineffective, as was the matrix sector-stack view (Figure 5.2C).

The tracklet sector and stack views (Figures 5.2A, 5.2B) were combined with the track views (Figure 5.3) to reduce redundancy. The combined views overlay tracks and tracklets, and distinguish between the two with both colour and line-style, encoding information in two visual channels. They also show underlying detector structure, and pair overviews and detail zoom views of a selected track. The components showing a *side view* of the detector are grouped as stack components, and those showing a *front view* are grouped as sector components.

We chose to add visualisations for raw ADC data recorded by the TRD in this iteration. Identifying an appropriate data source, and developing tooling to extract sample data, discussed in Appendix C, was a complex task. We considered using random noise as dummy data to reduce the implementation time required. The relationship between reconstructed and raw data was identified as important, however, and the effectiveness of visualisation options is best judged with realistic data.

This data is also quite complex, as each chamber is a 3D volume with the quantity of charge deposited at a given location as a fourth dimension, and the time of deposition as a fifth. The raw data display requirement thus involves a visualisation of this five-dimensional space[17]. We attempted to represent this data using two variants of flattened 2D histograms, with magnitude of charge encoded as a colour scale in both, and experimented with encoding time using animation.

We also attempted to show tracklets as the 2D planes they are, including the effect of the $2°$ tilt and uncertainty in z-position resolution. This required using more accurate detector geometry, which was applied to all existing components as well. Our attempts to represent this were unsuccessful due to scale-of-effect and level-of-detail issues (discussed in Appendix D), and the resulting components were not demonstrated to users.

The following user requests for additional functionality were deemed out of scope due to the effort required, and deferred to future work: scripting the event display; connecting to remote data sources; Run 3 time-frame visualisations; TRD operation status; data quality monitoring; and additional changes to the existing AliESD conversion task to simplify use.

5.2.3   New and updated components

We describe below both the refinements to existing components, and the development of new components, that were implemented based on the preceding interviews and feedback. The participatory design approach from iteration one was repeated here, and only the final components presented to users for feedback at the end of the iteration are discussed.

---

[17]Technically the z-dimension can be ignored, as the detector resolution in this direction is equal to the pad-length, so there is no variation in z for a given pad row.

(A) Tree-based track selection.

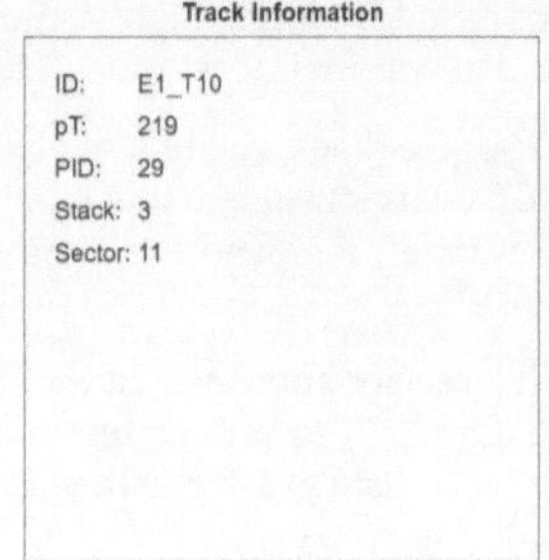

(B) Textual information for selected track.

FIGURE 5.6: The tree view shown in (A) is used to select an event or track. Additional information on the selected track is then displayed in the information box shown in (B).

5.2.3.1 Navigation and information views

The retained tree navigation component is functionally equivalent to the previous version in Figure 5.6A. Users can select a single event, or a track within an event, however tracklets were removed. When an event is directly selected all tracks within that event are assumed to be selected as well. Track labelling was changed to make explicit that these are tracks matched to TRD data, and now include the sector and stack that the track passes through in the description, as recorded in the ESD file. The default extraction task numbers tracks according to their index location within the ESD file, but this is customisable. Within an event tracks are ordered by their labels, as numerically close sectors are physically closer than numerically close stacks (which could be on opposite sides of the detector).

Figure 5.6B is a simple text box containing information related to the selected track, extracted from the source files, as requested by multiple users. This version displays a unique track identifier (ID), the reconstructed transverse momentum (p_T in GeV), the particle identification value calculated by the TRD (PID), and the stack and sector of the module the selected track passes through.

5.2.3.2 Stack components (side view)

The components in Figure 5.7 represent orthogonal projections of tracks and tracklets on to the zy-plane, using the same mapping as used in the iteration one component in Figure 5.3B. In this mapping each supermodule is individually rotated about the z-axis until its centre lies on the y-axis. All tracks passing through a supermodule are then rotated about the z-axis with the corresponding rotation, before being orthogonally projected onto the zy-plane.

Stacks are outlined in dark gray and numbered according to TRD convention; pad rows are outlined in light gray. Tracks are rendered as dashed blue lines – dark blue is used for selected tracks, light blue with lower opacity for other tracks, which allows overlaying of multiple tracks – and exhibit minimal curvature due to the magnetic field. The interaction point is shown at the bottom of the figure with an accurate relative distance to the TRD; the empty space between is occupied by inner detectors (TPC, ITS) which are not shown.

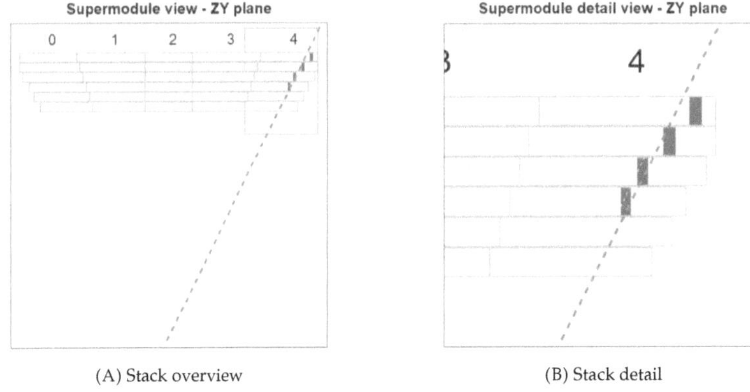

(A) Stack overview   (B) Stack detail

FIGURE 5.7: Overview and detail view of track and tracklet data
projected on to stacks.

There is inherent uncertainty in the z-position of a tracklet, which cannot be determined more granularly than within a single pad row. We chose therefore to highlight the entire pad row rectangles which contain a tracklet to illustrate this uncertainty. Dark red is used for tracklets matched to selected tracks, while light pink is used for other, non-selected tracklets – these colours differ in both hue and luminosity ensuring simple differentiation. This view clearly illustrates the 2D rectangular nature of fitted tracklets.

Figure 5.7B is a composite overview, showing tracklets and tracks for all 18 supermodules, normalised to a single ideal supermodule aligned with the zy-plane. A gray rectangular box is positioned over the stack containing the selected track. This box links the overview to the detail view in Figure 5.7A, which is a zoomed-in view of the contents of that box that makes the relationship between tracks and matching tracklets much easier to discern. The movement of the box between stacks is animated when the selected track changes, to better assist users in tracking the perspective and view change.

5.2.3.3 Sector components (front view)

The components in Figure 5.8 represent orthogonal projections of tracks and tracklets on to the xy-plane, the equivalent of looking down the beam-pipe toward the muon arm. The individual layers per supermodule are shown in gray together with the half-chamber dividing line, to aid orientation and spatial perception. Tracks are again rendered as dashed blue lines (dark for selected, light for others), and exhibit the expected curvature due to the magnetic field.

Tracklets are rendered as solid red lines (dark for selected, light for others) that are visually distinct from tracks. The tracklet line representation is effectively the side-on view of the 2D tracklet rectangle above. It accurately shows the location, slope and angle of the fitted tracklet, in contrast with the simpler black dot representation in Figure 5.3A. The effect of the $2°$ pad tilt is omitted in this view, as the effect is barely

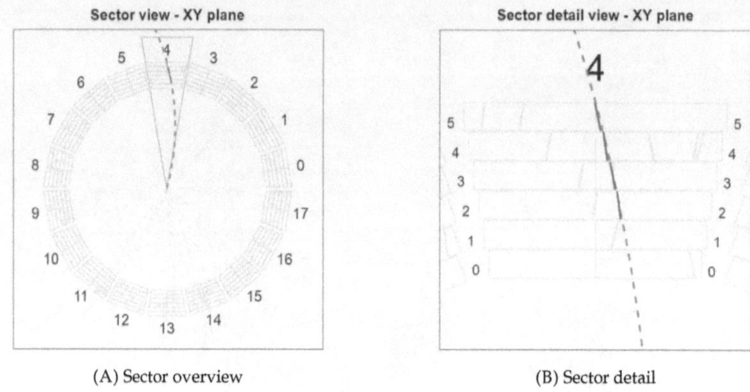

(A) Sector overview   (B) Sector detail

FIGURE 5.8: Overview and detail view of track and tracklet data
projected on to sectors.

noticeable[18]. The required projection of the 2D tracklet rectangle rotated about the y-axis would simply result in lines of slightly varying thickness that would be difficult to discern.

Figure 5.8A is an overview showing the interaction point and all 18 supermodules, numbered according to the TRD convention. The gray triangle in this figure highlights the sector that is then illustrated in greater detail in Figure 5.8B. The movement of the triangle between sectors when the selected track changes is again animated, and the zoom view visibly rotates to the appropriate sector, to assist users in maintaining context. The number of the displayed sector and numbering of individual layers according to convention help relate the displayed data to other work using this data.

5.2.3.4   Time-bin view

The static visualisation reproduced in Figure 3.9 combines two related views of raw ADC data from the TRD to illustrate how tracklets are reconstructed. We use this as a template for our time-bin component view in Figure 5.9.

This component shows up to six copies of the same visualisation, one per layer in a stack, in a scrollable fixed size window. For each of the six layers, if a tracklet has been matched to the selected track we use the matched tracklet position to select a pad row to display. If no tracklet is matched we hide all display elements for that layer. The subset of pad data displayed is a fixed range around the matched tracklet pad row and column.

The right side of this component shows the charge deposition on a range of pads (horizontal axis) over time (vertical axis) as a flat 2D histogram. The range of pads to display was heuristically assumed to be two pads either side of the tracklet midpoint. The time axis was partitioned into discrete time-bins, corresponding to how the ADC data is recorded, and a colour scale was used to indicate the quantity of charge deposited on a pad within a single time-bin. The colour scale was rendered in shades of green, in order to easily distinguish it from blue tracks and red tracklets – each major

---

[18]Discussed in Appendix D.

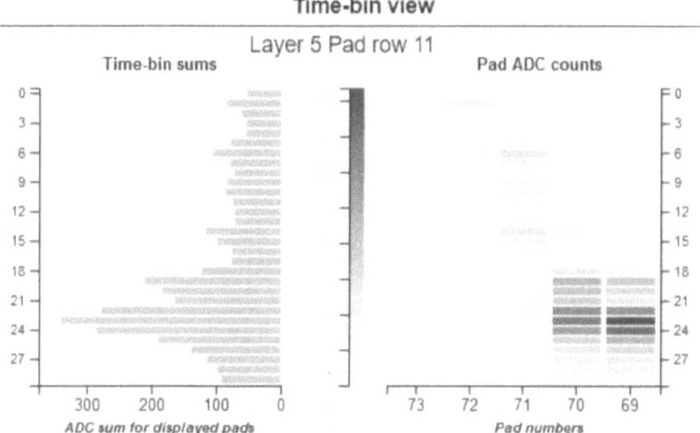

FIGURE 5.9: Presentation of raw ADC data in a time-bin view inspired by Figure 3.9.

data type is thus assigned a unique combination of primary colour, style[19] and saturation. The use of these three independent channels also helps mitigate issues associated with colour blindness.

The left side of this component is a 2D-histogram showing the total charge deposited per time-bin, across the range of pads displayed on the right. This view is comparable to the charge deposition over time curve in Figure 2.8, potentially allowing the general shape of the distribution to be compared against expectation. The combined view thus provides a link between the previous track and tracklet reconstructed data views, and raw ADC data that is the input to the reconstruction. We have omitted from this beta prototype the display of the y-coordinate, tracklet fitting, clusters, and Lorentz correction (all of which appear in the original static version) to limit the scope of work required.

When a track is selected, this component dynamically loads the ADC data for all the layers in the sector and stack that the track passes through. This data is stored in separate JSON files per stack-sector combination, and is extracted via the existing extraction task discussed in Appendix C.

5.2.3.5 Digits detail view

Figure 5.10 shows an alternative approach to displaying raw ADC data[20]. In this view we attempt to present all the raw ADC data for a single stack-sector combination, across all layers and available dimensions. Data used to reconstruct a tracklet is accumulated over a number of equally sized time-bins[21]. We attempt here to simultaneously show both the cumulative deposition over time on the left of the component, as well as the instantaneous deposition for a single time-bin on the right.

We experimented with several different approaches to presenting the data as discussed in Appendix D. The final version presented to users attempted a novel data

---

[19]Solid line, dashed line or filled rectangle.

[20]Colloquially referred to as digits data within the TRD group, hence the title.

[21]Typically 24 or 30 bins of 100ns each [5].

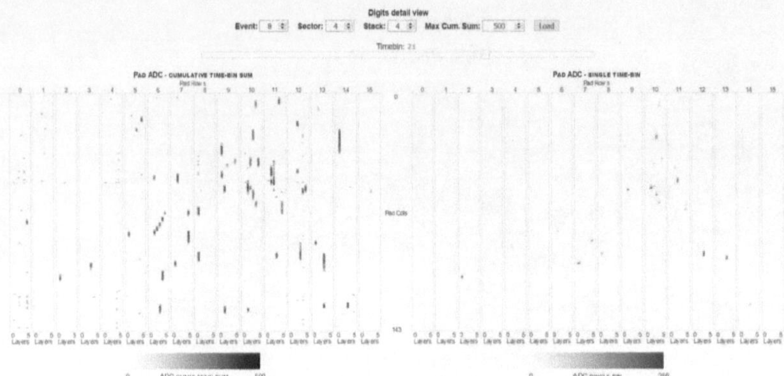

FIGURE 5.10: Cumulative and instantaneous charge deposition
in a stack, animated over time.

layout that placed related data close together, rather than mimic the physical layout.
A track generally interacts with a single pad row per layer, hence we have chosen to
group by numbered pad row along the horizontal axis, with a sub-ordering by layer
number. The vertical axis corresponds to individual pads within a pad row, and only
the start and end have been numbered to reduce visual clutter.

The shaded blocks in each pad row group represent the ADC value measured on that
pad. The colour scale uses green for the instantaneous and gray-scale for the cumulative view, chosen to match the time-bin component colours. Along the top of the
component is a slider that can be used to select a specific time-bin. The instantaneous
view on the right shows the ADC data measured in the single selected time-bin. The
cumulative view on the left shows the cumulative sum of all charge deposited from
the start of the event up until the selected time-bin.

Animation is used to automatically advance the slider from time-bin zero, with each
step displayed for one second. This allows the time information channel to be visualised, showing both how the charge deposition develops over time, and the contribution of each individual time-bin.

5.2.4   Implementation details

The iteration two **beta** prototype extends the existing HTML/SVG **alpha** prototype,
retaining the same technology stack and component based model. The JSON schema
specification was developed further to support ADC data, with two separate file
sources used per event: one for track and tracklet data; the other for ADC data from
a `TrdDigits` file[22]. JSON file data sizes were reduced by eliminating redundancy
and shortening field name lengths. All data was assumed to be available as static,
pre-generated files.

The potentially numerous tracks and tracklets that must be rendered were preemp-
tively accounted for by performance optimisations to SVG rendering and utilising the
HTML **Canvas** element for the *digits detail view* component. We expect data volumes

---

[22]Discussed further in Appendix C.

to be further managed by preparatory data cuts that focus on the areas of interest prior to export to JSON. These performance improvements ensure responsiveness is maintained as discussed in Section 3.3.3.

The complete iteration two beta prototype appears in Figure 5.11[23] and is available online at `https://datacartographer.com/alice-trd-event-display/v2/` [70].

Display considerations

We have used each of the primary colours (red, green, blue) as an information channel, along with shape perception, for different types of geometric data (tracklets, ADC data, tracks), and distinguish sub-classes with saturation and luminosity changes. Linear colour scales were used for magnitude perception of quantitative data.

Animated transitions, dynamic updating and selection highlighting were all used to assist users in navigating the interface and maintaining contextual awareness. Users can select events via the tree control, but further cross-filtering between components has not been implemented to reduce the complexity of implementation. Navigation is based on a data hierarchy where each element is defined by a hierarchy of keys (Run, Event, Track, Tracklet, Chamber) which, when selected, display a related series of values (Events, Tracks, Track path, Tracklet position, ADC count).

Temporal data display

Despite a single event developing over a period of time from the initial collision, we have mostly used simplified temporal semantics that flatten the time coordinate and display an aggregate view over the total duration of an event. This will need to be re-examined in light of the continuous data taking approach envisioned for Run 3. Two components attempted to include a richer temporal component through using time as a key (time-bin view and other rejected prototypes)[24], or encoding time via animation (digits view).

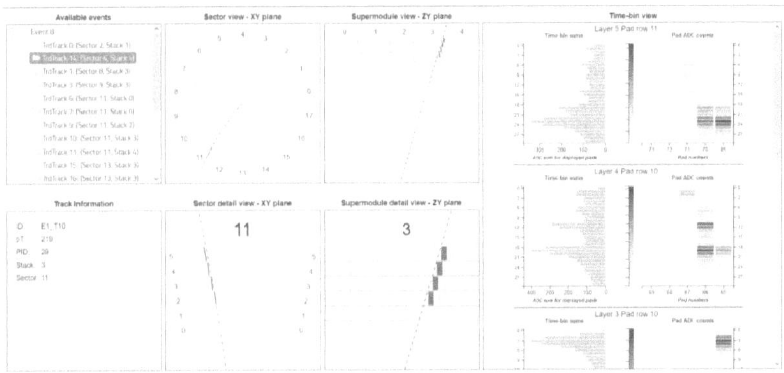

FIGURE 5.11: Screen capture of entire iteration one beta prototype as presented at ALICE TRD workshop.

---

[23]A larger, full-page version appears in Figure D.6.

[24]See Appendix D.

5.2.5   Feedback summary

The feedback from potential users was generally positive across the group presentations and hands-on sessions. Users found the revised components clear and easy to understand, navigation was logical, and different elements were visually distinct. One respondent enquired about the primary audience for this display, as "targeting different audiences is difficult and generally results in trade-offs". Members of ATLAS expressed interest in using this display for different detectors and experiments, validating the key design decisions to create a portable, web-based, easy-to-use display.

Requested changes

A number of improvements were suggested by participants, most notably the ability to see tracklets matched to some other, non-selected tracks. The demonstrated beta prototype only distinguishes between tracklets matched to the selected track and all other tracklets. The ability to display all tracks reconstructed from any set of detectors within ALICE, rather than only tracks that match TRD data, was also suggested as a way to broaden the potential usefulness of the display.

The centre of ALICE is not empty, yet Figures 5.7A and 5.8A show large empty spaces. Users suggested rendering the detectors there, particularly the TPC as it such a large component of ALICE. As in iteration one, many users commented that a 3D display would be useful to contextualise the 2D projections in the other views.

The time-bin view (Figure 5.9) was deemed useful for low-level data interrogation. Users did suggest that an overlay of the reconstructed tracklet on the ADC data, both with and without the Lorentz correction applied, would be useful in validating the reconstruction software being developed for Run 3 in O^2 .

Displaying additional track information in the text component, particularly related to collision parameters (type and energy) and the triggers involved, was highlighted as a useful link between the visual display and underlying data.

Issues

Several issues were identified by participants and required addressing in the next iteration. The selected range of displayed pads in the time-bin view (Figure 5.9) appeared incorrect and inconsistent. The apparent center of charge deposition did not always agree with the central pad chosen for display.

The digits view component (Figure 5.10) proved complex and confusing to users, and the unusual layout was difficult to parse. The visuals and animation were deemed interesting, but no clear use was identified by any participant. The related lack of a legend to explain what each colour and symbol represented was also highlighted by several users.

The geometry of displayed stacks, sectors, and layers were updated for this iteration, following user feedback. However the supermodule geometry was again identified as being incorrect, and the use of the singular "Supermodule" obscured the fact that data from all 18 supermodules was actually presented in a single normalised view. Users also enquired if it was possible to show pad boundaries within a layer, and that the units of momentum be added to the text information component.

Visual queries

Without being prompted, several users asked and attempted to answer many of the questions highlighted in the initial interviews[25] during the demonstrations. This suggests that even in its current state, the display is effective in finding patterns or anomalies that can then be further investigated through traditional analysis.

5.2.6 Reflections on the process

The opportunity to visit CERN in-person, view the physical detector, and interact with the scientists who work on it was very valuable in understanding the domain and refining the prototype to better align with reality.

The demonstration of effective dynamic visualisations opened the door to other applications of data visualisation that might not previously have been considered. It also created a virtuous cycle as initial prototypes elicit feedback related to successes or shortcomings, which drives improvements in the visual representation, which in turn elicits further feedback.

As highlighted in feedback, the detector geometry used was thought to be more accurate relative to iteration one, but mistakes were made during extraction which were identified by users. This correlates with our experience of abstract designs being difficult to improve, as scientists tend to focus on the details only once the high level representation is acceptable. We also found that having familiarity with the domain simplified discussions with participants, and resulted in higher quality critical feedback.

5.3 Final prototype: Iteration three

5.3.1 CERN Open Days

CERN Open Days is an outreach exercise conducted approximately every five years during a Long Shutdown. Over the weekend of 14-15 September 2019, 80 000 members of the public visited CERN sites in Meyrin, Prévessin, and each of the experiments[26].

We were invited to include our final prototype display as part of the TRD stand in the ALICE display at Point 2. This opportunity provided an ideal platform to engage with a cross-section of the public and garner their input on our prototyped design, as an initial secondary aim of the study was to understand how event displays could best aid communication with the general public.

Figure 5.12 shows the display stand with the prototype displayed to the left. The bright natural light in the venue required modifications to the existing prototype to ensure maximum visibility. Given the different expectations of a public audience, we also chose to add the frequently requested 3D view component, as well as cleaning up several visual aspects of the components. These changes are documented below.

---

[25]Discussed in Section 5.1.1.

[26]ALICE, ATLAS, CMS, and LHCb.

FIGURE 5.12: Setup of the CERN open days stand. The final prototype event display is on the monitor to the far left, next to a poster summarising the TRD detector. Supermodule number 7 forms the backdrop, with a cut-out to show the internal electronics.

5.3.2 Final design

The existing beta prototype from iteration two was retained and updated based on all previous user feedback as well as the specific requirements for the public display. The components presented below are the final versions displayed at CERN Open Days 2019.

3D component

The biggest change from the previous beta prototype is the addition of an interactive 3D display of the TRD detector supermodules, with tracks and tracklets overlaid – various views of this are shown in Figures 5.13 and 5.14. The primary aim of this component is to provide an overview of the detector and aid orientation of users who are unfamiliar with the TRD. The 2D rectangular nature of tracklets is also clearly illustrated in this 3D view.

The default behaviour constantly rotates the camera view in the horizontal plane, without user interaction, to limit the occlusion of important data. Users are also able to manually rotate, zoom and pan the camera view with the mouse, allowing data to be viewed from any angle. A series of buttons along the bottom of the component allow further customisation; detector elements, tracks and tracklets can all be independently hidden and the automatic rotation of the camera can be disabled.

Selection behaviour

We added the display of both TRD and ESD[27] tracks, and added the ability to better distinguish matched and unmatched tracks in response to user feedback. This necessitated changes to the colour schemes and display behaviour for different selection

---

[27] Tracks matched by other detectors within ALICE, but not the TRD.

(A) No event selected   (B) Event selected   (C) Detector hidden

FIGURE 5.13: 3D component display options when an event is selected.

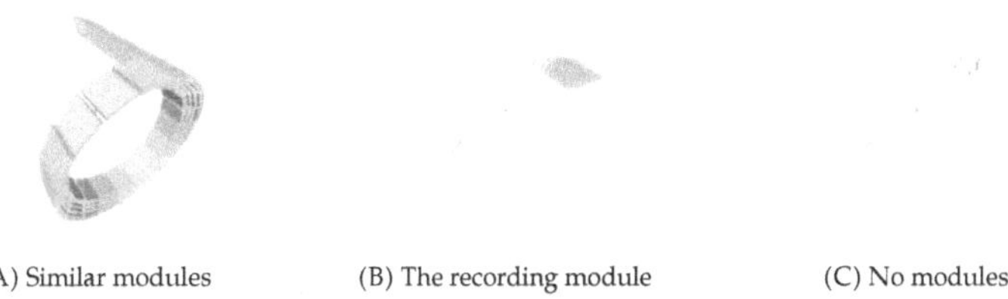

(A) Similar modules   (B) The recording module   (C) No modules

FIGURE 5.14: 3D component display options when a track is selected.

states. Figure 5.18 illustrates the colour and style encodings used in this final prototype, and the legend element introduced to clarify this for users.

ESD tracks are coloured light blue with low opacity, selected TRD tracks dark blue, and non-selected TRD tracks purple. This allows users to appreciate the relative abundance of ESD vs TRD tracks, whilst still being able to visually distinguish selected tracks. We apply the convention here that when an event is selected, all TRD tracks and tracklets within that event are presumed selected as well. In a similar vein, selected tracklets are coloured red, and non-selected tracklets yellow. We additionally colour tracklets that are matched to a track that is not the selected track orange. This is intended to help answer the tracklet visual queries mentioned above, by allowing users to better differentiate between matched and unmatched tracklets.

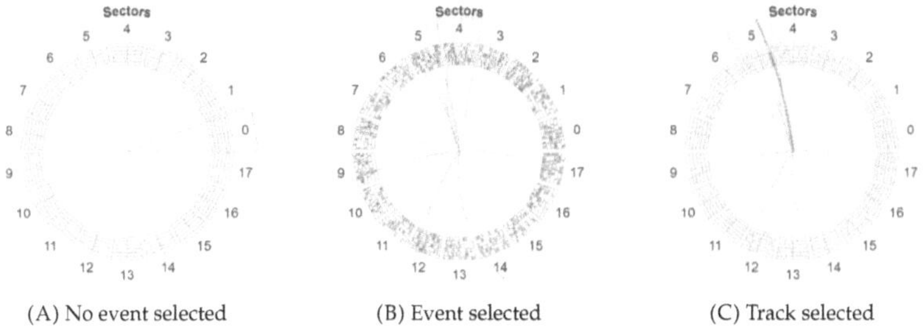

(A) No event selected   (B) Event selected   (C) Track selected

FIGURE 5.15: Sector component display for each possible selection state.

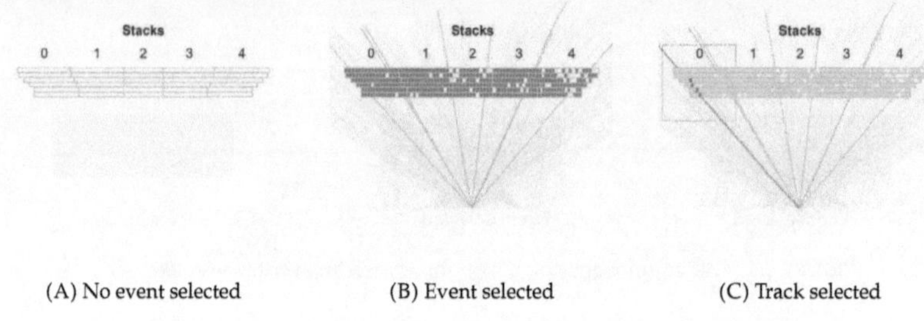

<table>
<tr><td style="text-align:center">(A) No event selected</td><td style="text-align:center">(B) Event selected</td><td style="text-align:center">(C) Track selected</td></tr>
</table>

FIGURE 5.16: Supermodule component display for each possible selection state.

This selection behaviour is illustrated in Figures 5.15 and 5.16 for the sector and supermodule views respectively. The same colour scheme is also applied to the 3D component as illustrated in Figures 5.13 and 5.14. The zoomed detail sector and supermodule views are hidden when only an event is selected, but are shown in the same colour scheme when a track is selected, as in Figure 5.17. The detail view additionally only shows tracks and tracklets recorded in the single supermodule that the selected track passes through, to reduce visual clutter. Finally the text information component highlights important track parameters in blue, continuing the colour scheme.

The 3D component has three display mode toggles for detector elements, which helps spatially locate the relationship between a track and the stack it transits. Figure 5.13 shows the three possible display modes when only an event is selected. Figure 5.14 shows the corresponding modes when a track is selected, allowing a user to: identify with which sector ring and supermodule the track interacts (Fig 5.14A); view only the stack through which the particle transits (Fig 5.14B); or hide the detector entirely to better view the relationship between the selected track and all other tracks (Fig 5.14C).

Display improvements

The digits display component (Fig 5.10) was removed based on user feedback and the time-bin view was rescaled and moved below the 3D display. This creates a functional layout where the primary 2D track and tracklet projections are the central focus of the display, with navigation, context and raw data on the periphery.

The time-bin view in Figure 5.19 has been extended to include the reconstructed tracklet before and after Lorentz drift slope correction[28], differentiated by line style to maintain the red selected tracklet colour encoding. The range of pads displayed for a given track was fixed, and the y-coordinate axis from Figure 3.9 was reintroduced.

The empty spaces in the stack and sector overviews were filled with a low opacity toroidal projection representing the TPC, which contextualises the TRD with respect to the largest detector in ALICE without distracting from the actual data. The correct pad and pad row dimensions, including irregular end sizes, were applied by re-extracting the geometry for the detector as a whole.

---

[28]See Section 2.2.4.

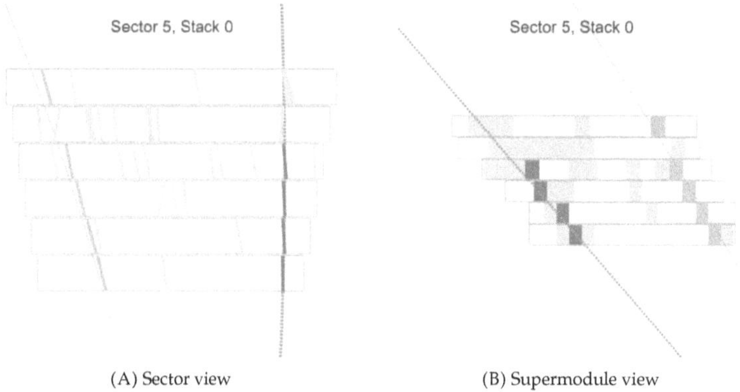

(A) Sector view   (B) Supermodule view

FIGURE 5.17: Detail component views when a track is selected.

Finally the textual information component was enhanced to display information in full sentences (helpful to novice audiences), and important data values are highlighted in blue, as shown in Figure 5.20.

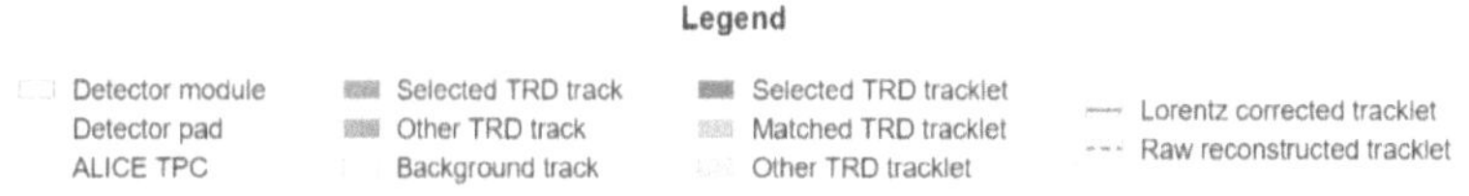

FIGURE 5.18: Legend detailing encoding of tracks, tracklets and ADC values for final prototype.

5.3.3 Implementation details

A responsive design approach [40] was adopted and tested on all four major browsers (Firefox, Chrome, Safari, Edge) to ensure the event display final prototype was truly cross-platform. The display was also verified to be functional on mobile devices, but the user experience was far from ideal.

The 3D display component was implemented using **WebGL**, an industry standard for displaying 3D graphics on the web [11], and **three.js** [84], a JavaScript library for rendering 3D graphics in a web browser.

The JSON data formats for reconstructed and raw data were both finalised in this iteration, and the data size for the ADC JSON data was reduced by only generating data for modules with selectable tracks. The final version of the event display, as presented at CERN Open Days, is shown in Figure 5.21 and a full page version in Figure D.7. An online version is available at `https://datacartographer.com/alice-trd-event-display/` [71].

Time-bin sums
Layer 3 Padrow 3
LocalY pad coordinate (cm)
Time bin
ADC sum for displayed pads
Pad ADC Count
Pad number
Time bin

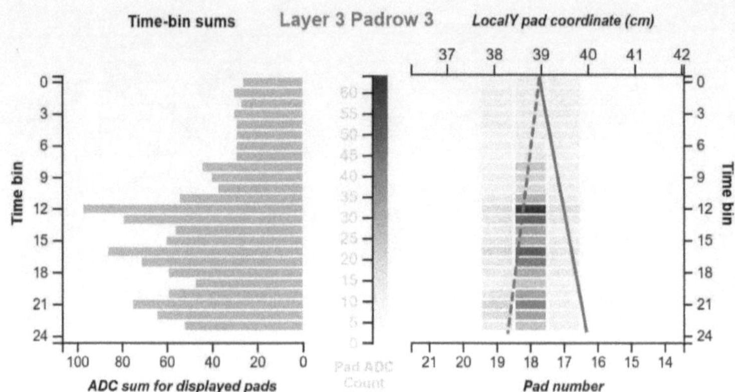

FIGURE 5.19: Time-bin component with visual display changes.

Track information
Event
Collision between a proton and a lead nucleus at an energy of 3.5 TeV
Track
Trd track E98_T100 traverses Sector 5, Stack 0 of the TRD with a transverse momentum of 2.14512 GeV. The calculated PID value of 17 indicates this is likely a pion track.
Tracklets
4 of the 1425 tracklets detected by the TRD have been matched to this track.
Triggers
17 high-level triggers fired for this event: 0VBA, 0VBC, 0TVX, 0VIR, 0STG, 0UBA, 0UBC, 0VHM, 0OM2, 0SH1, 0SH2, 0V0M, 0BPA, 0BPC, 1HCO, 1ZED, 1ZMD

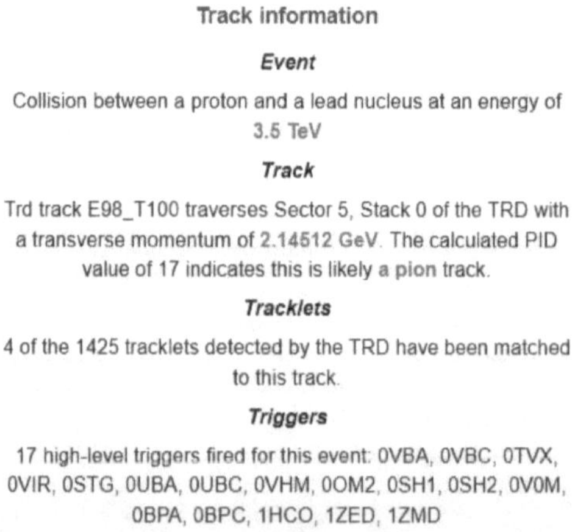

FIGURE 5.20: Text information component with visual display changes.

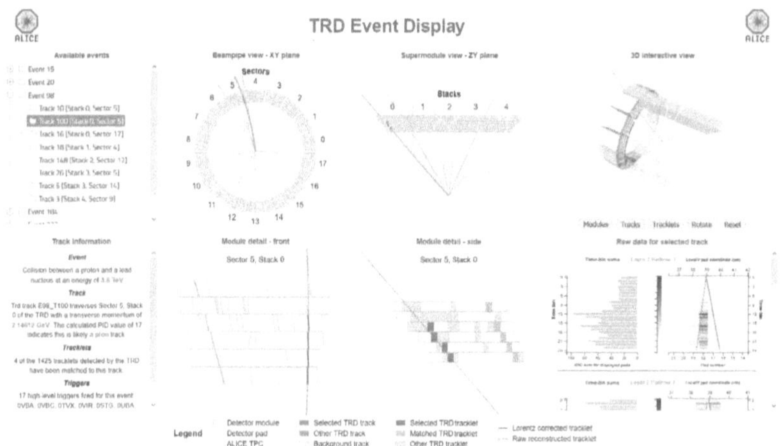

FIGURE 5.21: Full view of the final prototype display.

5.3.3.1 Public feedback

The feedback from the public at CERN Open Days, both indirect through discussion and direct through anonymous written comments, was overall very positive – the most common documented comments were variations on "the display helps non-scientists understand how the detector works".

The 3D component was most frequently commented on, with most respondents finding it helpful in establishing context and understanding what a track was, and how it related to the displayed supermodule in front of them. The 2D projections were useful in demonstrating specific effects, such as pair-production and the charge dependence of curvature direction in a magnetic field. Several members of other experiments at CERN who visited the ALICE display also expressed interest in utilising the display for their data.

The parallel display of information across components of both raw and reconstructed data was specifically highlighted as being useful in "understanding the different stages of data reconstruction"[29]. Some feedback did suggest that it would be useful to be able to expand particular components, as the relatively small screen made it difficult to discern details at a distance. This could be resolved by using a large, high resolution screen for future public displays.

One respondent specifically requested "Detector material highlighting in 3D display" as they were interested in what materials were used to construct the detector. This was the first feedback of this type, and suggests that a display geared toward the engineering details of the detector could be of use[30].

We found, and the public specifically commented, that a story telling approach to describe the functioning of the detector and the process of reconstruction, using the

---

[29]This comment was from a member of the public with some particle physics knowledge, but no previous exposure to ALICE.

[30]A virtual reality tour of the detector using Computer Aided Design (CAD) software was part of the ALICE exhibition, which might have prompted this request.

Date	Description	Designation
2019-06-29	Presentation of **alpha** prototype to combined ALICE and ATLAS group at the University of Cape Town.	D1
2019-07-10	Presentation of **beta** prototype at TRD mini-week in Frankfurt.	D2
2019-07-10	Hands-on demo session of **beta** prototype held after TRD mini-week in Frankfurt.	D3
2019-09-12	Interactive session with ALICE PhD student at CERN to setup and use **final** prototype.	D4
2019-09-15	Inclusion of **final** prototype in CERN open days.	D5
2019-09-30	Demonstration of **final** prototype to 3rd year physics students at the University of Cape Town.	D6
2019-12-05	Use of **final** prototype by ALICE Physics Honours student at UCT investigating tracking efficiency.	D7
2020-01-24	Use of **final** prototype by post-doctoral researcher at CERN to visualise simulated data from O^2 .	D8

TABLE 5.1: Timeline of all presentations and demonstrations of event display prototypes.

event display as a animated visual aid, had the most positive reception and worked best for a general, non-specialist public audience. This suggests there is great scope for future work that adopts this approach in communicating otherwise intimidating physics concepts and results to the general public.

5.3.4 Case study demonstrations

During the course of this work, we were presented with several opportunities where the event display was utilised by physicists within the TRD group as part of their ongoing research. These demonstrations, and participant responses, validated the creation of this new event display and the functionality it provides. The full list of demonstrations appears in Table 5.1.

An interactive session (D3) was held after the presentation of the iteration two beta prototype to the TRD group in Frankfurt. This prototype was used to discuss the difference between TRD and ESD tracks, and the relationship between tracks and the stack and sector they transited through. This ultimately led to the inclusion of ESD tracks in the final prototype, and validated the representation and accuracy of the existing track display.

Across several discussions with the PDE, potential issues with the relationship between tracklets and data were identified. This led to adoption of a modified version of this event display to assist in debugging the tracklet reconstruction code being overhauled in preparation for Run 3, as well as help learn about sources of error.

An ALICE PhD student at CERN approached us after seeing a demonstration to discuss use of the display with their own data that they were trying to understand. They were able to setup, configure and use the display (D4) with their own data with minimal guidance, and commented that the final prototype immediately proved useful and easy to use.

An ALICE Physics Honours student at UCT, investigating the tracking efficiency of conversion photons, used the display to visually investigate why some electrons (and thus the photons they come from) were being rejected even though they met certain trigger criteria (D7). They found the display of trigger ids in the text information component of particular use when viewing their data. The final prototype display was also used in a successful hands-on introductory tutorial session with 3rd year physics students at UCT (D6).

An ALICE post doctoral research associate used the final prototype display (D8) to present work-in-progress results of Run 3 simulation data at a TRD weekly meeting. This was done using only the JSON specification in Appendix C and validates the choice of a simple intermediate format (as the Run 3 data format had not yet been finalised) and the low barrier to entry posed by the display.

5.4 Personal reflections

This study was begun with a high degree of *impostor syndrome* [27] induced trepidation, owing to the highly technical and specialised nature of the domain. It was therefore a great relief to find that all the scientists who participated were very helpful and forthcoming. It was nevertheless difficult to avoid the necessity of acquiring sufficient domain specific knowledge to ask the right questions and appropriately interpret responses. This cross-domain work was enthusiastically received, as scientists don't often have capacity to develop visualisations, but find great value in well constructed, easy-to-use examples. We did find that specialists tended to focus on their area of expertise, so we needed to canvas a range of potential users to adequately establish generic requirements.

There was a notable difference in reception and areas of focus between the primary (scientists) and secondary (general public) audiences. Scientists tended to be interested in asking questions about data and finding anomalies, given their familiarity with the domain. We also found an unmet need for visualisations of more technical considerations both within the detector and its resultant data.

The general public and novice users sought instead a general understanding of aims and outcomes of experimental particle physics. A small subset of those were also interested in the operation of the detector, and the high level interpretation of recorded data. We found that visualisation for communication is often best accompanied by audio narration or explanation, as telling a story is more engaging than reading dry facts and figures.

During the design study process we found that free-form, semi-structured interviews worked very well, allowing diversions into areas that would otherwise not have been considered. Many of the prepared questions from Appendix A were not applicable to particular respondents, or required knowledge they did not have, and the ability to steer conversation towards areas of actual interest to them proved useful. This suggests that overly structured questionnaires are sometimes inappropriate in scientific contexts with complex domains.

The combination of interviews supplemented by rapid prototypes for discussion proved very fruitful, as we were able to quickly integrate ideas from users as well as limiting time wasted on ineffective or impractical ideas. The difficulty of designing for multiple audiences led us to initially focus on one, however it transpired that well designed components can still provide value for different audiences, as long as they are focused and don't try to show too much detail.

The application of visualisation principles was very helpful in guiding development, as they led to clarity of communication. Aesthetic considerations were definitely valuable in attracting interest to the final prototype (as evidenced in public feedback), but the functionality provided was ultimately essential to both audiences deriving value and understanding from the final prototype.

6 Conclusions

In this work we present the results of a design study for an event display focused on raw and reconstructed data from the TRD in ALICE at CERN. We have used a participatory design approach, interviewing particle physicists individually, and documenting and synthesising the interview responses to create a component-based design. This design was implemented as a series of prototypes across three iterations, and each prototype was critically evaluated in both individual and group settings, with the feedback used to inform and refine the design in the following iteration.

We have created a portable, web-based display, dedicated to the TRD and its data, that is also extensible to other detectors and experiments. It includes the ability to display and link both raw and reconstructed data in a single interface, which can assist in the validation of O^2 reconstruction routines being developed for Run 3. The components of this display were designed using visualisation best practices to ensure information is effectively and clearly communicated to users via multiple visual information channels, always putting function ahead of form.

We document both the utilisation of these prototypes by physicists, in several case study demonstrations, to understand data related to research questions, and the experience of designing with scientists. In so doing we establish and validate a successful process template for user-centric design of visualisations in the field of high-energy particle physics. We hope these observations can prove useful in developing future data visualisation displays, both within the field of physics as well as other related scientific disciplines.

The final prototype was demonstrated to the public during the CERN Open Days event, where it was well received. The valuable public feedback we received validated many design choices that were initially made with a specialist rather than general audience in mind, evidence that well designed visualisations can work for multiple audiences.

During this process we gained a working knowledge of the CERN software ecosystem, and ROOT in particular, which we used to implement a generic software tool that extracts both raw and reconstructed data from AliESD ROOT files into JSON. The JSON data interchange format we have defined is used to allow our event display to consume data from many potential sources, and future web-based event displays can build on this as a common standard.

We have documented our personal reflections on the experience of conducting a design study in a very technical scientific domain for an audience of expert users. Gaining a sufficient understanding of the domain and associated terminology was essential to being able to ask the right questions and propose appropriate solutions. Individual interviews helped uncover the varied detailed needs of users, while group sessions allowed ideas and critiques to build on each other in an effective feedback loop. The use of visualisation was unanimously welcomed as a useful tool to help understand complex data. Rapid, iterative prototyping in conjunction with collaborative, user-centric design produced an effective, functional prototype. We are very proud that this prototype has been actively used by scientists in the field, as well as for outreach to the general public.

6.1 Future work

6.1.1 Integration with ALICE software ecosystem

The current implementation includes an automated, documented software tool to extract raw and reconstructed data into JSON, which was developed for Run 2 data accessed through AliRoot. The current upgrade to the O^2 framework requires this tool to be extended to support Run3 data, both simulated and detector recorded. It would also be useful to upgrade this tool to output information from a specific sub-region of the detector, and to appropriately handle cases where only a subset of data types (track, tracklet, or ADC) is available. The visualisations presented here could also be extended to display **quality control** (QC) information, either for offline validation or as part of online monitoring in the ALICE control room during LHC runs. Finally, the visualisation designs and choices used in this implementation could be included as part of the ongoing work to create a generic, ROOT-based event display for ALICE in O^2 .

6.1.2 Higher data volumes

The current implementation assumes a relatively small subset of data is displayed, in order to maintain responsiveness in potentially low-resource browser environments. As previously mentioned, Run 3 is likely to significantly increase the data volumes recorded by the TRD and ALICE compared to previous runs. While the interface elements we have designed should also be applicable to dense data representations, additional visual information channels may need to be exploited for maximum effectiveness. Technologies and techniques could also be investigated to optimise the display within the same browser-based infrastructure. Server side pre-rendering of graphics and data compression could be effective in reducing client-side resource requirements. Where the current implementation statically loads most information at startup, dynamic loading of data from online storage could be very effective in managing load times and resource utilisation.

6.1.3 Interactive visualisations

User interaction in the current implementation is limited to event/track selection from the tree control, and moving the camera in the 3D control. Enhancements could investigate the use of cross-filtering, where clicking on elements of any component would contextually update the current filter criteria[1] and reduce the displayed data accordingly. The ability to customise component layout and capabilities would be very valuable, as would functionality to annotate elements of interest – these annotations could then be exported or added to the original data source. Remote control of the event display through scripting, and synchronised views of the same data across multiple browsers, were other common requests from users that could be implemented in future versions.

---

[1] For example, click on a sector to show only tracks and tracklets in that sector.

6.1.4   Exploration of animation

The data recorded by the TRD is complex and often time dependent in multiple ways (the track left by a particle over a period of time, the gradual signal build-up on a pad, the pile-up of signals due to collisions occurring in rapid succession). It would be interesting to explore the use of animation to encode this temporal data. In particular, although the approaches attempted in Section 5.2.3.5 and Appendix D were ultimately unsuccessful, we believe there is potential in revisiting and refining them to better illustrate the charge deposition on pads over time.

6.1.5   Outreach

The CERN hosted *Master Classes* introduce high-school students to high-energy particle physics through hands-on activities. The portable, no-setup nature of the current implementation lends itself well to being rapidly deployed on multiple computers as part of such outreach activities, and the visual nature of the display is well suited to novice users (as validated by public feedback during CERN Open Days). This controlled environment is well suited to conducting a study evaluating the effectiveness of such use, which could then inform future general science outreach activities.

A Initial interview structure

Semi-structured interviews were conducted as a series of leading open-ended questions to provide guidance to interviewees in navigating the problem space. The interview template is presented and discussed below:

- *What is your background?*

  This question aims to understand the interviewee's level of expertise within the general field of high energy particle physics, in order to contextualise the responses to later questions.

- *What is your role in ALICE?*

  This question establishes the interviewee's level of expertise within the ALICE experiment, and leads in to follow-up questions on related to familiarity with the TRD and its data.

- *What is your primary interest in the TRD data?*

  This question uncovers different high-level use cases to potentially incorporate into the design. The intent is to understand how the data is used, regardless of potential for visualisation, which can be useful in helping the researcher identify possible visualisations not previously considered.

- *Are there any specific areas where you think visualisation might be useful?*

  This question builds upon the previous by uncovering how the interviewee thinks about data and the visualisation thereof. It can be followed by discussion about examples or types of visualisation that the interviewee is not familiar with, which can in turn lead to new areas of use.

- *Is there specific functionality that you would require in a new visualisation tool?*

  This question directly relates to the interviewee's experience with existing visualisation tools, and guides the discussion toward combining the answers from the previous questions into actionable items that can be prototyped.

- *What research questions would this functionality help answer?*

  This question helps understand how the previous answer relates to the real-world needs of the interviewee, and the forced mental connection can uncover other potential needs related to the same research area.

- *What are the shortcomings of the existing tooling, specifically as relates to the previous question?*

  This is the first negative question, and allows the interviewee to express dissatisfaction with the status-quo. The intention here is to understand whether the frustration is related to: desired features that have not been implemented; features that have been implemented poorly but have the potential to be improved; or desired functionality that requires a completely new approach or solution.

- *What features of existing tooling must be retained?*

  This question ensures that in the process of creating a new design, the most valuable features of existing work is retained. Discussion can also branch into how these must-have features can be improved.

- *How would you envision interfacing with other analysis code?*

  This question relates specifically to the difficulty of applying visualisations to the large data volumes and complex analysis common to the field, as uncovered in the foundation phase reviews.

B Summary of variables

Symbol	Definition	Value or (usual) units
c	speed of light in a vacuum	$299\,792\,458\,\mathrm{m/s}$
$m_e c^2$	electron mass $\times c^2$	$0.510\,998\,946\,1\,\mathrm{MeV}$
β	v/c ratio of particle speed to speed of light	
γ	$1/\sqrt{1-\beta^2}$ Lorentz factor	
z	charge number of incident particle	
Z	atomic number of absorber	
A	atomic mass of absorber	$\mathrm{g\,mol^{-1}}$
K	$4\pi N A r_e^2 m_e c^2$ Coefficient for dE/dx	$0.307\,075\,\mathrm{MeV\,mol^{-1}\,cm^2}$
I	mean excitation energy	eV
W	energy transfer to an electron in a single collision	MeV
W_{max}	Maximum possible energy transfer to an electron in a single collision	MeV
$\delta(\beta\gamma)$	density effect correction to ionization energy loss	

TABLE B.1: Summary of physical variables and constants
used in equations [83]

C Data wrangling

C.1 JSON Input Format

The **final** prototype visualisation uses a web-standard JSON [39] data format, opti-mised for the display of TRD-specific information. Overview data is initially loaded for a set of events and tracks that are displayed in the tree view control. This data format is described in Section C. Detailed pad-level information is loaded on demand when a track is selected - this data format is described in Section C. Due to the poten-tially large data volumes that can be generated, JSON field names have been abbrevi-ated where possible, generally to three characters.

The following conventions are used in the listings below:

- Each listing shows the first level field of a JSON object. Where a field is an object (*{...}*) or array of objects (*[...]*), the expected structure is shown in subsequent listings.

- Each member is followed by a comment indicating: its purpose; the valid range (if one exists); and whether the field is required (*req*) or optional (*opt*).

- The listing captions begin with a name in all-caps (e.g. *EVENT*). This is the JSON objects type name, and is used to reference it in other listings.

Event data

At the highest level, the display operates on a series of events. The corresponding JSON that is expected consists of a simple list of EVENT records.

LISTING C.1: TOP-LEVEL INPUT - List of EVENT records to display

```
1  [ {...}, {...}, ... ]
```

Each EVENT record corresponds to a single collision event. Each record must have an *id* field whose value is a unique identifier of the form "E*" where * is a number.

LISTING C.2: EVENT - data for all tracks and tracklets in a single trig-gered collision event

```
1  {
2    "id":     "E5",       // Event ID, req
3    "i":      {...},      // Track information, opt
4    "tracks": [...],      // List of TRACK objects, req
5    "trklts": [...]       // List of TRACKLET objects, req
6  }
```

An EVENT record may contain an optional *i* field with additional information stored in standard JSON object key-value-pair format. An EVENT record must have both a *tracks* field with a list of TRACK objects (C.3), and a *trklts* field with a list of TRACKLET objects (C.4).

LISTING C.3: TRACK - data for a single track

```
1  {
2    "id":    "E5_T1",      // Track ID, req
3    "stk":   0,            // Stack, 0-4, req
4    "sec":   0,            // Sector, 0-17, req
5    "typ":   "Trd",        // Type of track: { Trd, Esd }, req
6    "i":     {...},        // Track information, opt
7    "path":  [...],        // List of POINT, req
8    "tlids":               // List of matched tracklet ids, req
9      ["E5_L1", ...]
10 }
```

Each TRACK record represents a single reconstructed track within the parent EVENT. Each record must have an *id* field whose value is a unique identifier of the form "E*_T*" where * is a number, and the first part of the identifier corresponds to the parent EVENT *id*. The *stk* and *sec* fields are required, and represent the stack and sector where this track first enters the TRD. A record may contain an optional *i* field with additional information stored in standard JSON object key-value-pair format.

The *typ* field is required, and may have one of two values: "Trd", which indicates a TRD track with matched TRD tracklets; or "Esd", which indicates a reconstructed track with no matched TRD tracklets. The *tlids* is a required field, and is a list of strings representing TRACKLET *ids* (C.4), though the list can be empty. If the *typ* field is "Trd", the list should contain the *ids* of all matched tracklets.

The *path* field is required, and is a list of 3-dimensional POINT objects (C.5). These points should be sampled from the reconstructed track trajectory, and are rendered as a sequence of connected line segments. A record may contain an optional *i* field with additional information stored in standard JSON object key-value-pair format.

LISTING C.4: TRACKLET - data for a single TRD tracklet

```
1  {
2    "id":    "E5_L1",      // Tracklet ID, req
3    "stk":   0,            // Stack, 0-4, req
4    "sec":   0,            // Sector, 0-17, req
5    "lyr":   0,            // Layer, 0-5, req
6    "row":   0,            // Pad row, 0-15, req
7    "trk":   "E5_T1",      // Id of matched track, opt
8    "lY":    0.0,          // LocalY coordinate, req
9    "dyDx":  0.0           // DyDx, slope of tracklet, req
10 }
```

Each TRACKLET record represents a single tracklet measured by the TRD within the parent EVENT. Each record must have an *id* field whose value is a unique identifier of the form "E*_L*" where * is a number, and the first part of the identifier corresponds to the parent EVENT *id*. The *trk* field is optional, and if populated should contain the *id* of the corresponding TRACK that this TRACKLET was matched to during reconstruction.

The *stk*, *sec*, *lyr* and *row* fields are required, and represent the stack, sector, layer and pad row where this tracklet was measured by the TRD. The *lY* and *dyDx* fields are required and represent the reconstructed localY and slope of the tracklet.

LISTING C.5: POINT - 3D space-point in standard TRD coordinate frame

```
1  {
2    "x": 0,              // x-coordinate, cm, req
3    "y": 0,              // y-coordinate, cm, req
4    "z": 0               // z-coordinate, cm, req
5  }
```

Each POINT record represents a point in 3-dimensional space, within the standard TRD coordinate frame. All coordinates are in units of centimetres and are required.

Pad data

Within the display, certain views visualise the raw ADC/Digit data recorded by the TRD. These views are populated whenever a track is selected, using the EVENT *id* and TRACK *stk* and *sec* values to dynamically load the corresponding JSON data, if it exists. This data is stored as a DIGITS (C.6) object, which has a required *evid* field that corresponds to the selected EVENT *id*. The record also contains a required *lyrs* field, which is a list of LAYER objects (C.7) that represent the raw data for each layer within a module.

LISTING C.6: DIGITS - Raw data for a single event in a single module

```
1  {
2    "evid": "E5",        // Event ID, req
3    "lyrs": [...]        // List of LAYER objects, req
4  }
```

LAYER objects are only required for layers in which data was recorded. Each LAYER must have a *lyr* field containing the layer number as per TRD numbering convention. It must also have a *pads* field, which is a list of PAD objects (C.8) which contain the raw data for individual pads.

LISTING C.7: LAYER - Data for a single layer

```
1  {
2    "lyr": 0,            // Layer index, 0-5, req
3    "pads": [...]        // List of PAD objects, req
4  }
```

A PAD object must have required fields *row* and *col* which correspond to the pad row and pad column of the PAD. It must also have required field *tbins*, which is an array of integer values, representing the ADC count for each time-bin, starting at time-bin 0. It is not necessary to include PAD objects where there is no data, or where all time-bins are 0.

LISTING C.8: PAD - Data for a single pad across all time-bins

```
1  {
2    "row":    0,              // Pad row position, 0-15, req
3    "col":    0,              // Pad column position, 0-143, req
4    "tbins": [0, ...]         // ADC counts per time-bin, uint, req
5  }
```

C.2 Run 2 data extraction

We developed a standard AliAnalysis task[1], in parallel to the various event display iteration prototypes. This task extracted real recorded data from ROOT data files stored within the ALICE Environment (ALIEN) grid framework [73], and created corresponding JSON data files that were read by the event display.

The initial task for iteration one extracted event, track and tracklet data from *AliESDs.ROOT* files, and required understanding the API and storage formats, as well as projecting tracks and tracklets from local coordinate systems to a global coordinate space. This functionality was extended in iteration two to export ESD tracks and track-tracklet matches from the *AliESDs.ROOT* file, as well as raw data from *TRD.Digits.ROOT* files. The raw ADC data had to be decoded using [57] as a reference, and the JSON data was stored in individual files per stack-sector combination. The potential size of extracted JSON data necessitated this choice so that individual files could be loaded on demand, rather than loaded upfront as is the case with event and track data. The final code for this task is available online at: `https://github.com/samperumal/msc-cpp/`.

---

[1]Following the tutorial at: `https://alice-doc.github.io/alice-analysis-tutorial/`

D Implementation details

D.1 Detector geometry

The accurate display of TRD detector elements in both 2D and 3D displays requires an accurate 3D definition of the geometry of each element of the detector to be displayed. This geometry was manually extracted from technical sources ([5], [49], [57]) into a spreadsheet and then converted into JSON using python scripts[1]. The resulting JSON description file is imported by the event display, and appropriately projected by each component for rendering. This allows for support of multiple detectors and experiments by simply modifying the JSON file contents.

The use of idealised geometry from literature has downsides, as the actual detector geometry changes after installation due to effects such as deformation of the space-frame holding the detector, variances in size during manufacturing, and adjustments required during routine operation. The correct approach would be to use calibrated geometry data from files which are updated at the start of every run, as these account for run-dependent variations.

The manual extraction of geometry was complicated by several subtle variations between otherwise identical elements, which required several rounds of refinements to accurately capture the true ideal geometry. The number of pad rows per stack differs, with stack 2 containing 12 pad rows, and all other stacks (0-1, 3-4) containing 16 pad rows. The width of an individual pad is also dependent on both the stack and layer within which the pad is located: pad lengths are minimal in stack 2 and maximal in stacks 0 and 4; pad widths are minimal in layer 0 and maximal in layer 5. One final aspect is that there is a constant space between pads and pad rows, except before the first pad and pad row, and after the last pad and pad row.

D.2 Tracklets

The transition from the simple representations of tracklets in iteration one to the more correct line/plane representation in later iterations required gaining an in-depth familiarity with the storage format of tracklets, and the associated coordinate systems. Tracklets are stored with an integer value for the stack, sector, layer and pad row in which they occur. Their position within a pad row is stored as floating point value (localY) that indicates the distance from the center of the layer at which the tracklet enters. The slope of the tracklet (dyDx) is then used to determine the point at which the particle exited the drift region of the chamber. These values must then be rotated into the global ALICE coordinate system for rendering.

The tracklet fitting function provides a position resolution of 140 nm along the pad row, but is unable to resolve the z-position with any more granularity than the length

---

[1]Source code available here: `https://github.com/samperumal/msc-python/`

of a pad. After discussions with the PDE on how best to illustrate this uncertainty, we settled on the solutions shown previously, where a tracklet is actually a 2D plane, the projection of which is simply a rectangle the length of the corresponding pad in the z-direction. This length varies slightly along with the previously mentioned pad length dependence on layer and stack, but is very difficult to distinguish at the scales this event display operates at.

D.3 Pad rotations

The pads within a pad row are tilted by $2°$, in alternating directions per layer, in order to increase the z-resolution of fitted tracklets, as illustrated in Figure 3.8B. We attempted several approaches to including this in the iteration two prototype, but the 2D projections of the rotated tracklet plane were indistinguishable from the simple line representation of the tracklet with no rotation – the scale of the effect was simply too small to notice at the zoom levels used in the prototype.

D.4 Digits view

As documented in Section 5.2.3.5, we attempted several representionns of raw ADC data. Our first attempt appears in Figure D.1, which is a simple 2D histogram showing the total quantity of charge deposited per pad on a pad row within a layer, for all layers within a selected stack. Figure D.2 is a more polished version of the same idea, with labelled axes, a colour scheme based on a sequential linear gradient of a colour not used elsewhere in the interface, and the nesting of pad rows within layers clearly illustrated. In both cases data is shown using pad/pad row coordinates, rather than distances. This results in equal sized layers that are easier to compare, but the spatial relationship between vertically adjacent pads is lost and overlaying tracks is no longer possible.

Figure D.3 shows a slightly more sophisticated attempt that pairs the same 2D histograms with the ability to select a single layer. This then brings up a series of line graphs, one per pad row, that show the charge deposition over that pad row. This view makes it easy to see spikes where potential tracklets might be located, but with no other way to correlate this with other data like tracks and tracklets.

The final attempt was the animated consolidated layer view, discussed previously and shown in a much larger view in Figure D.4. As mentioned, though interesting, this version proved too complicated to interpret to be of immediate use to users. Our hope is that with the experience gained from this work and additional interactions with users, this can be turned into a good example of the value animation can bring to event displays.

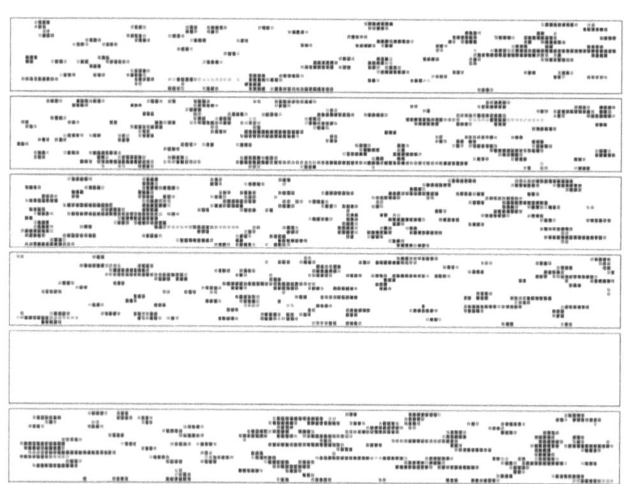

FIGURE D.1: Simple 2D histogram of cumulative charge deposition per layer over a single stack.

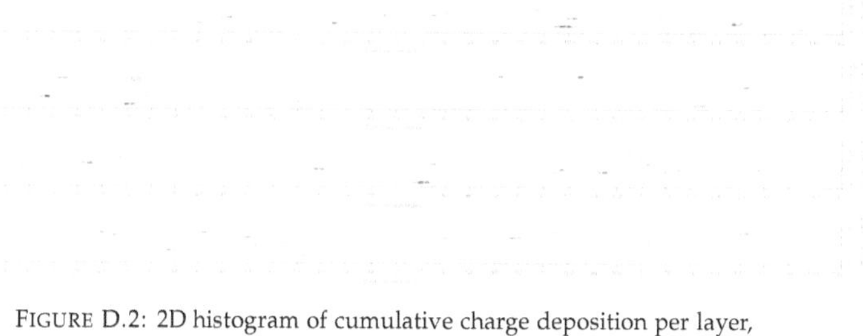

FIGURE D.2: 2D histogram of cumulative charge deposition per layer, over a single stack with labelled axes.

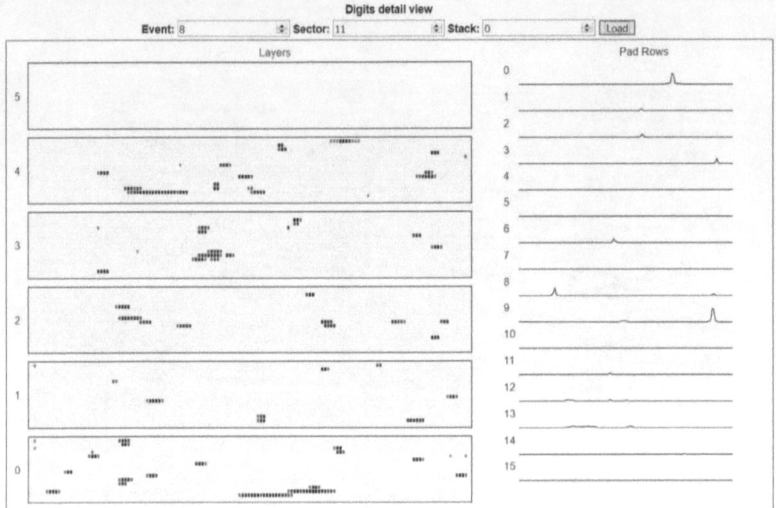

FIGURE D.3: 2D histogram of cumulative charge deposition over a single stack with pad-row line graphs for the selected layer.

D.5 UCT TRD summer school

As part of the UCT High Energy Physics summer school program, we put together a prototype web interface to replicate and extend the TRD monitoring functionality of SMMon (*SuperModule Monitor*). This interface allows users to modify and monitor the running state and setup of one or more TRD modules, in a way that can be widely shared with no pre-requisites. The code for this interface is available here: `https://github.com/samperumal/uct-trd-ui`

D.6 Complete prototype views

Full page views of each iteration prototype as presented to users appear below for iterations one (Figure D.5), two (Figure D.6) and three (Figure D.7) to better display all details of that particular prototype.

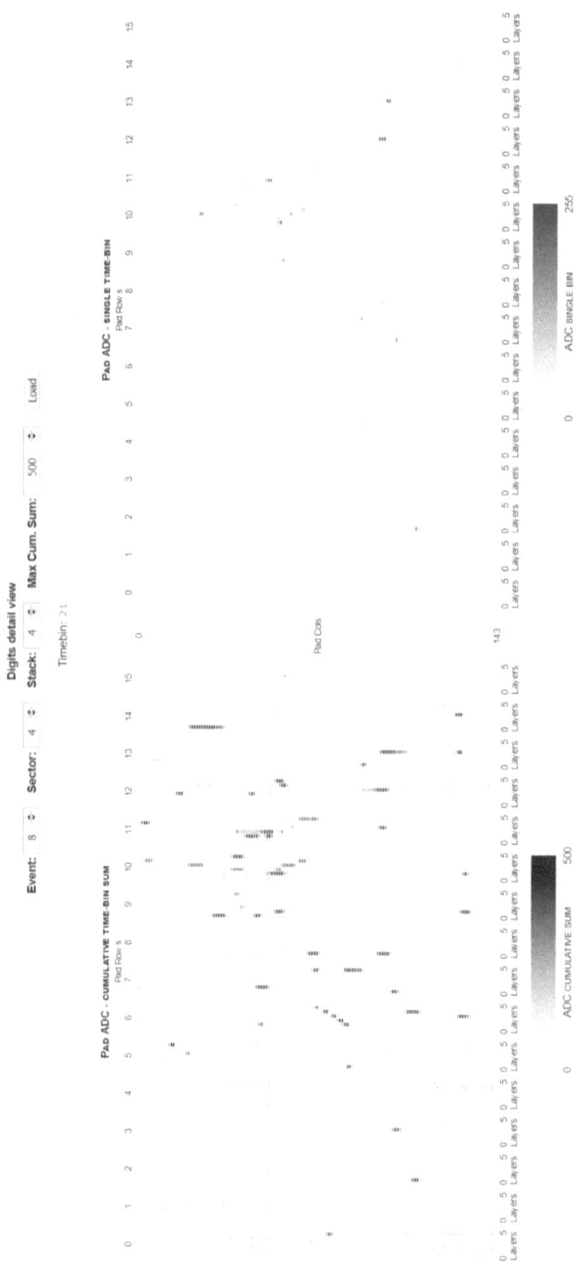

FIGURE D.4: Cumulative and instantaneous charge deposition in a stack, animated over time.

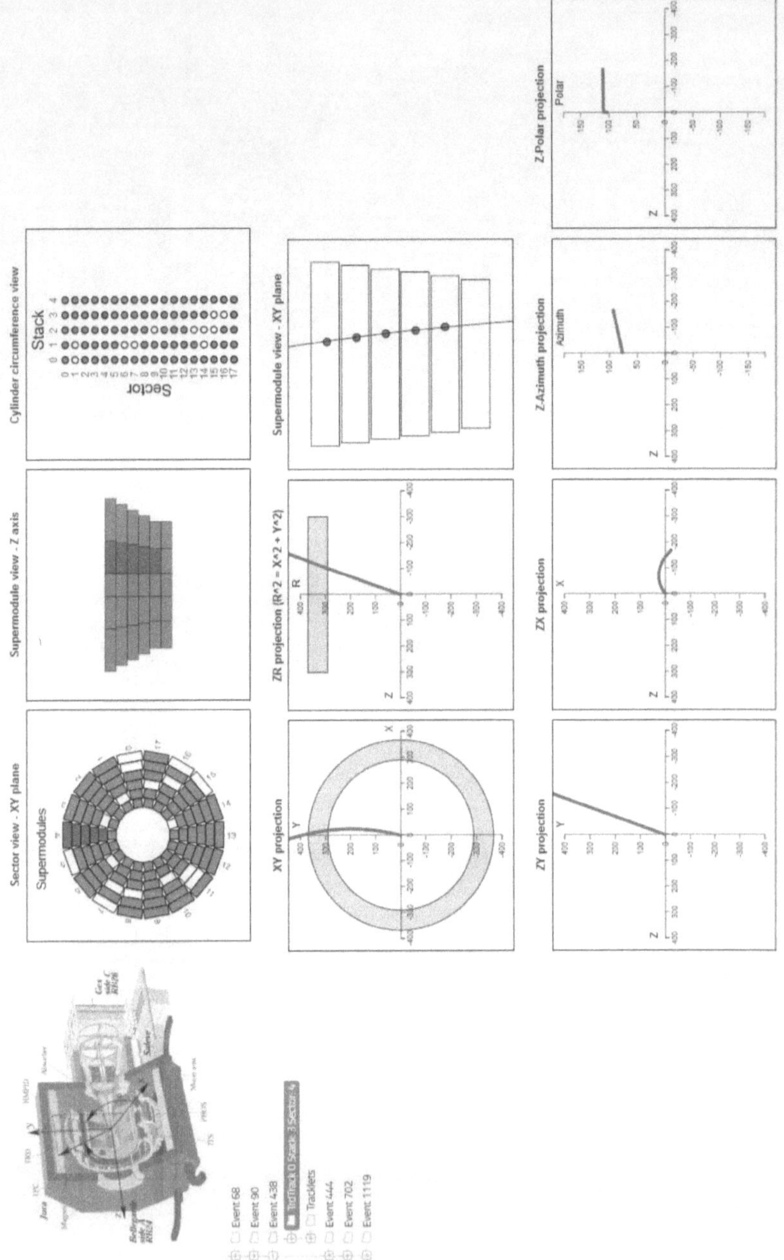

FIGURE D.5: Large view: alpha prototype.

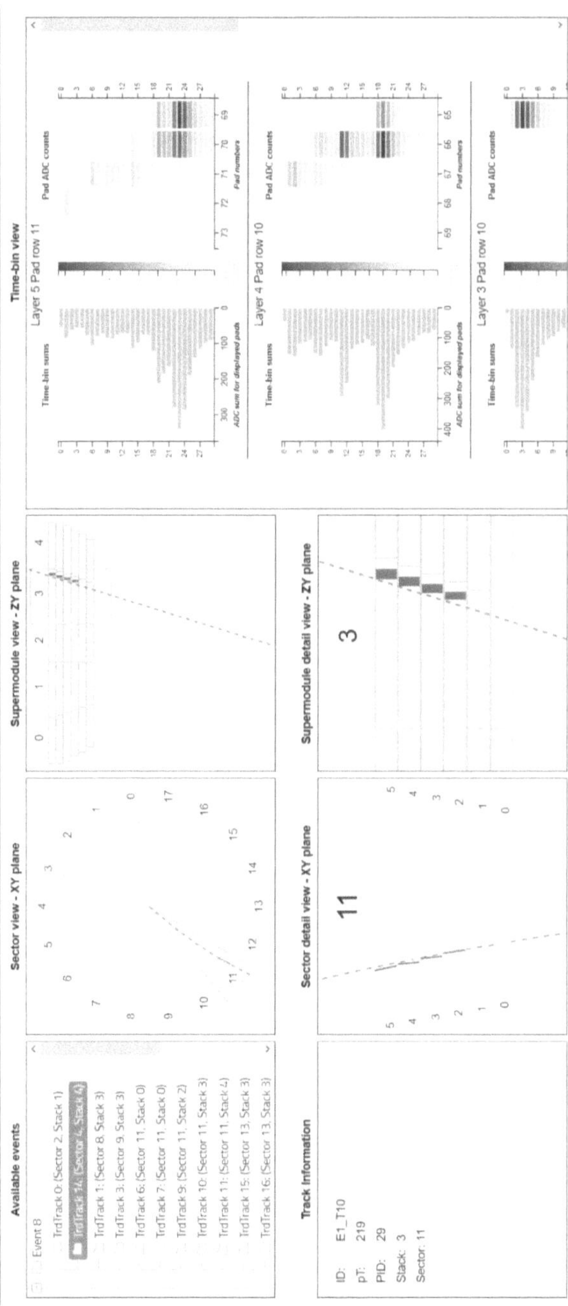

FIGURE D.6: Large view: beta prototype.

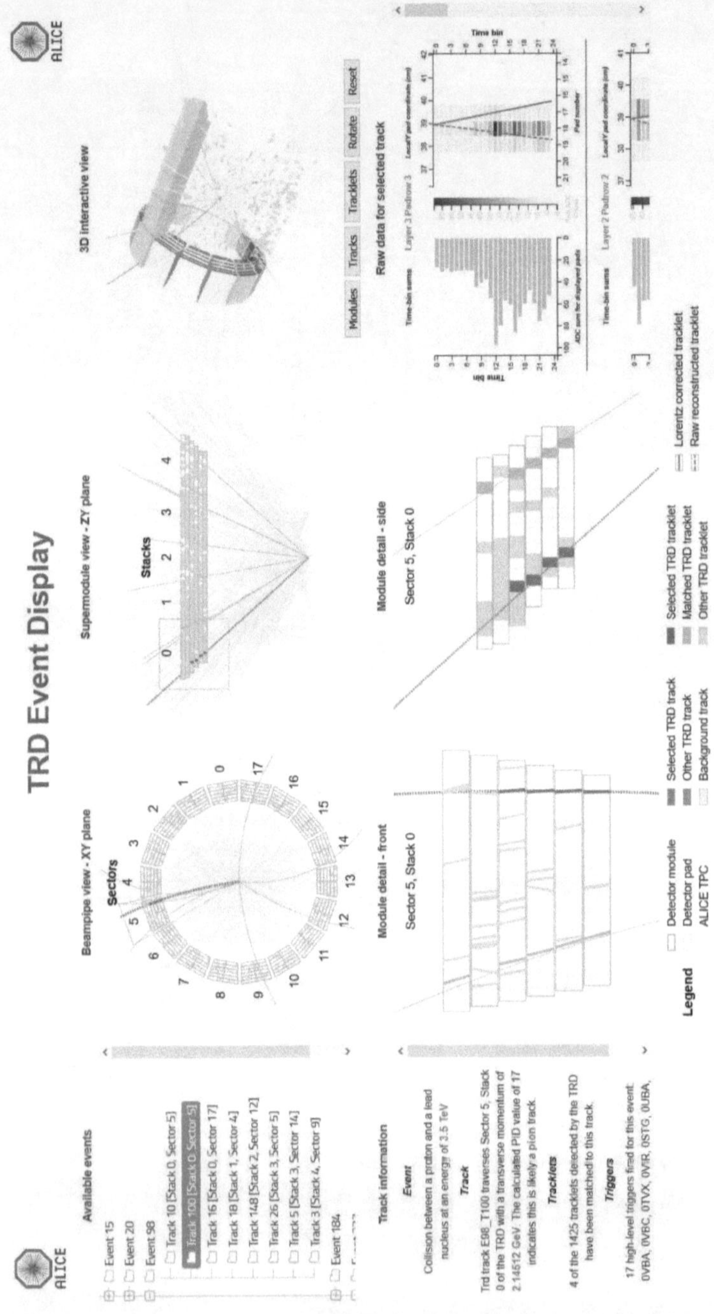

FIGURE D.7: Large view: final prototype.